D0908221

Professional Communication in Engineering

Palgrave Studies in Professional and Organizational Discourse

Titles include:

Keith Richards
LANGUAGE AND PROFESSIONAL IDENTITY

H.E. Sales
PROFESSIONAL COMMUNICATION IN ENGINEERING

Forthcoming titles include:

Edward Johnson & Mark Garner
OPERATIONAL COMMUNICATION

Cecilia E. Ford
WOMEN'S TALK IN THE PROFESSIONAL WORKPLACE

Palgrave Studies in Professional and Organizational Discourse
Series Standing Order ISBN 0–230–50648–8
(*outside North America only*)

You can receive future titles in this series as they are published by placing a standing order. Please contact your bookseller or, in case of difficulty, write to us at the address below with your name and address, the title of the series and the ISBN quoted above.

Customer Services Department, Macmillan Distribution Ltd, Houndmills, Basingstoke, Hampshire RG21 6XS, England

Professional Communication in Engineering

H.E. Sales

First published 2006 by
PALGRAVE MACMILLAN
Houndmills, Basingstoke, Hampshire RG21 6XS and
175 Fifth Avenue, New York, N.Y. 10010
Companies and representatives throughout the world

PALGRAVE MACMILLAN is the global academic imprint of the Palgrave Macmillan division of St. Martin's Press, LLC and of Palgrave Macmillan Ltd. Macmillan® is a registered trademark in the United States, United Kingdom and other countries. Palgrave is a registered trademark in the European Union and other countries.

ISBN-13: 978–1–4039–4806–9 hardback
ISBN-10: 1–4039–4806–2 hardback

This book is printed on paper suitable for recycling and made from fully managed and sustained forest sources.

A catalogue record for this book is available from the British Library.

Library of Congress Cataloging-in-Publication Data
Sales, H.E. (Hazel E.), 1948–
    Professional Communication in Engineering / by H.E. Sales.
        p.   cm.
    Includes bibliographical references and index.
    ISBN 1–4039–4806–2 (cloth)
    1. Communication of technical information.   2. Engineering.   I. Title.
    T10.5.S26 2007
    620.001 4—dc22   2006046067

10   9   8   7   6   5   4   3   2   1
15   14   13   12   11   10   09   08   07   06

Printed and bound in Great Britain by
Antony Rowe Ltd, Chippenham and Eastbourne

To all engineers, and those who aspire to be.

To all migrants and those who aspire to be

Contents

List of Figures   xi

List of Tables   xiii

Acknowledgements   xiv

1 The Engineers   **1**
  1.1 Introduction: technological creativity   1
    1.1.1 Engineering is essentially a creative profession 1
  1.2 Why they became engineers   3
  1.3 Engineers have to write a lot   6
  1.4 The customer   8
  1.5 Types of engineer   9
    1.5.1 Snap-shot of an engineering work-force 9
  1.6 Creativity versus restriction   14
  1.7 Communication skills: engineers' aspirations   19
  1.8 Final comment on the discourse community and aims
     of this book   21

2 Engineering Practices and Procedures   **23**
  2.1 Engineering procedures   23
    2.1.1 The burden of responsibility   23
    2.1.2 Engineering methodologies: the importance of
        procedures   24
    2.1.3 Engineers' attempts to control writing
        behaviour   27
  2.2 Writing procedures and guidelines   30
    2.2.1 Written, and then often ignored   30
    2.2.2 Simplified English   32
    2.2.3 Information Technology Security Evaluation
        Criteria (ITSEC)   35
    2.2.4 Customer-imposed writing constraints   36
  2.3 Collaborative writing practices   39
    2.3.1 What do engineers write with other people? 39

2.4 Risk-taking and the textual straw man 41
   2.4.1 Collaborative writing case study 44
   2.4.2 Evaluating the straw man: engineers' opinions 51
2.5 Summary 53

3 The Engineering Product **55**
3.1 Introduction: setting the scene 55
3.2 The product life-cycle 56
   3.2.1 Knowledge accrual in engineering design 56
   3.2.2 Temporal considerations in product design 60
3.3 A textual perspective of the product life-cycle 64
   3.3.1 Thresholds and stages in product development 65
3.4 Summary 69

4 Engineering Texts **71**
4.1 Introduction: engineers' view of texts 71
   4.1.1 Formal versus informal 72
   4.1.2 Ephemeral versus 'lasting' texts 73
4.2 The documents engineers write 75
   4.2.1 Engineers don't write anything they
      don't value 75
   4.2.2 The product is the focus 77
   4.2.3 The main text types 81
   4.2.4 The problem documents 84
4.3 Special features of engineers' language 86
   4.3.1 Technical description is far from simple 86
   4.3.2 Complex simplicity 88
4.4 Summary 90

5 Engineering Specifications and Requirements **92**
5.1 Introduction: delivering on a promise 92
5.2 What are specifications and requirements? 93
   5.2.1 Clarifying terms 93
   5.2.2 Specifications are the textual bedrock of
      engineering 95
   5.2.3 A glut of guidelines, but a dearth of information 96
5.3 The Customer Requirement 98
   5.3.1 Cardinal point specifications: customer's
      wishes and ideas 99
5.4 Control of the design process: the need to manage
   change 103
   5.4.1 A recipe for success (or disaster) 103

5.5	Categories of specification	109
	5.5.1 Hierarchies in requirements	109
5.6	A special language: engineers have devised their own rules	111
	5.6.1 What engineers don't like about 'English'	111
	5.6.2 The special case of modal verbs	113
	5.6.3 Monitoring requirements in computer databases	116
	5.6.4 A specification document case study	119
5.7	Concluding observations	121
6	**The Bid Process and Persuasion**	**122**
6.1	The bid process	122
	6.1.1 Proposal writing consumes huge financial resources	123
	6.1.2 Types of proposal	124
	6.1.3 Readers and writers of engineering proposals	126
	6.1.4 The writers and bid preparation	128
	6.1.5 The Customer Requirement: a catalyst for proposal writing	130
	6.1.6 Economic and social impact of winning (or losing)	134
6.2	Persuasion	136
	6.2.1 Engineers versus marketing colleagues	136
	6.2.2 Attempting to pin down the notion 'persuasion'	136
	6.2.3 Engineers' ambivalence towards overtly persuasive language	139
6.3	Engineers' attempts at restrained persuasion	142
	6.3.1 Restrained persuasion: Example 1	142
	6.3.2 Restrained persuasion: Example 2	145
	6.3.3 Restrained persuasion: Example 3 – composing text for a generic proposal	146
6.4	Concluding observations	152
7	**The Presentation of Engineering Proposals**	**153**
7.1	Introduction: textual cosmetics	153
7.2	Physical features: size, formats, and outline structure	154
	7.2.1 The textual 'extent' of proposals	154
7.3	Significant parts of the proposal	159
	7.3.1 An overview of each section	159
	7.3.2 Generic outline structure of proposals	160
	7.3.3 The front cover: engineers' use of visuals	163
	7.3.4 Proprietary page: laying claim to design rights	172
	7.3.5 Acronyms unlimited	174

7.4   Oral presentations of proposals: a case study   178
    7.4.1   Feedback on oral presentation   182
7.5   Conclusion   186

**8   Engineering Proposals: Discourse and Information
Structure   187**
8.1   Introduction   187
    8.1.1   Proposals persuade   187
    8.1.2   'Selling point' versus 'benefit'   189
8.2   Guidance on proposal writing: a historical perspective 189
    8.2.1   Up to 1980: the popularity of report writing   190
    8.2.2   Post-1980: proposals receive scant attention   193
8.3   A hierarchy of discourse functions   196
    8.3.1   Macro-level discourse functions   196
    8.3.2   Eight key discourse topics   196
8.4   Themes   201
    8.4.1   Themes according to Ellis (1997)   202
    8.4.2   Themes according to Stross (1990)   202
8.5   The identification of proposal components (PCs)   206
    8.5.1   The search for an analytical framework   206
    8.5.2   Towards a more topic-focused description   208
8.6   Conclusion: reverse-engineered text   212

9   Executive Summaries   214
9.1   Introduction: what are executive summaries?   214
    9.1.1   Purpose of the executive summary   214
9.2   Structure, information content, and presentation of
    executive summaries   218
    9.2.1   The summary as a notion   218
    9.2.2   A neglected genre   219
    9.2.3   Persuasion through selling points: discourse
        functions   220
    9.2.4   Examples of executive summaries   226
    9.2.5   Executive summaries and proposals: a structural
        comparison   231
    9.2.6   Patterns of information structure   234
9.3   Summary   240

References   241

Index   246

List of Figures

1.1	Time engineers spend on writing	7
1.2	Work expended by different engineer categories	11
1.3	Engineer's view of documentation he has to produce	15
2.1	Using a straw man to construct a manual	46
3.1	A cyclic representation of the design process	58
3.2	A product life-spiral	59
3.3	Simplified product life-cycle: three main phases	61
3.4	A text-oriented view of the product life-cycle	62
3.5	Ideal post-delivery scenario	68
4.1	An engineer's depiction of text categories	73
4.2	Problematic writing	85
5.1	Extracts from a customer requirement document	99
5.2	A hierarchy of technical specification	110
6.1	Unexpected email initiates proposal writing	131
6.2	Persuasive strategy continuum (Houp and Pearsall 1980: 141)	137
6.3	The 'New spin' text (first page)	148
7.1	A 'route map' of a large proposal	156
7.2	Paula from Tech Pubs carrying a proposal	157
7.3	Pictorial representation of proposal sections	159
7.4	A front cover with the pencil picture	164
7.5	Proposal cover for a bespoke product intended for a destroyer	167
7.6	Proposal cover for a product designed for the Tornado aircraft	168
7.7	Proposal cover for product with wide application	171
7.8	Proprietary statement: laying claim to design rights	173
7.9	A glossary in a technical proposal	176
8.1	Houp and Pearsall's list of items for proposals (1980: 345)	191
8.2	Macro-level discourse functions and topics of technical proposals	197
8.3	Persuasion strategies in engineering proposals	201
8.4	A selection of Stross's themes (Stross 1990)	203
8.5	Taxonomy of proposal components (PCs) under four main 'what' categories	211
9.1	Outline generic structure of the executive summary	221

9.2 Generic structure of the executive summary: component
      realisations   222
9.3 Executive summary, example 1 – traditional plain format   227
9.4 Executive summary, example 2 – presented to show key
      benefits   228
9.5 Executive summary, example 3 – mimicking popular
      publications   230
9.6 Executive summary, example 4 – in the style of a 'glossy'
      magazine   232
9.7 Main PC categories in executive summaries   233
9.8 PC patterns in technical proposals   235
9.9 PC patterns in executive summaries   236

List of Tables

7.1   The textual 'extent' of a fairly large proposal   154
7.2   Typical breakdown of a proposal   158
7.3   Generic outline structure of engineering proposals   160

Acknowledgements

I would like to thank a British engineering company for enabling me to work amongst engineers over a prolonged period, showing its commitment to intellectual endeavour and openness to different ideas by offering me, an arts-oriented researcher, access to its science-oriented community. The company has provided work facilities and access to texts and other data in a friendly work environment, for which I am extremely grateful.

Colleagues have been unstinting in their support, in particular, Malcolm Maciver, Roger Wearne, and Simon Isserlis, who held the fort during the writing of this book. I would like to thank Dave Edmondson and John Kirkman for their help and being generous with their thoughts about engineering specifications, Tony Wright for his advice, and my institution, the College of St Mark and St John, for its sponsorship and encouragement. Tony Dudley-Evans, Tim Johns, John Sinclair, Florence Davies, and Michael Hoey, through their inspirational teaching when they were at the University of Birmingham, provided me with much to mull over and influenced the ideas in this book.

Jo Smart, Lucy Ellis, Steve Oldfield, Anthea Brooks and Alan Malvern read various drafts with care and insight. Their comments have contributed hugely to this book, and I am exceedingly grateful to them.

John Pullen, the technical illustrator, whose talent and wizardry with the computer elevated my humble drawings to art forms, merits special mention and thanks.

Special note: Most of the names of people, companies, projects, products and locations in this book are pseudonyms.

1
The Engineers

1.1 Introduction: technological creativity

1.1.1 Engineering is essentially a creative profession

Ask almost any engineer why he or she chose to be an engineer and the reply may reveal something akin to an ideology that underpins their working lives: a high-minded aspiration to be useful to the community at large. They may not wear these ideals on their sleeves, but probe beneath the surface and you will find an altruistic streak. Engineers tend to have a strong sense of purpose, believing they have a contribution to make to society, and knowing that, through designing a myriad of things we use in everyday life (components, gadgets, software, all sorts of machines, buildings, bridges, and even the shoes we wear), their work affects nearly every aspect of human activity. So, engineers see themselves as being essentially creative, and working towards some kind of solution that has been asked for. This is a distinctive feature of their work: engineers have to be creative to order. The customer looms large on the horizon of any engineering workplace. Solutions need to be designed and produced for customers, the more ingenious the better.

It could be said then that engineers see their roles at work as being constructive and productive, which gives them a strong sense of raison d'être, and an altruistically motivated one at that. Like aspiring novelists, musicians and poets, who can see or hear the physical fruits of their compositions, engineers believe their contributions have some kind of measurable, physical 'presence' that contributes towards helping others in their endeavours. This book is based on a 6-year study of engineers working on products for the aerospace, defence, and automotive industries (Sales 2002). It pays particular attention to engineers working on design, although it is informed by those from other areas

1

within engineering. The study has an ethnographic basis, having been influenced by anthropological and sociolinguistic research methodo- logy (Saville-Troike 1989). It involved working within an engineering community for extended periods, and examining the texts, both spoken and written, that the engineers produced in the course of their work. The aim of the study was to learn from the engineers themselves about the communication tasks they perform at work, the texts (and language) they produce, and their views about them. It is a 'grounded' study, inspired by the work of others who have used similar ethnographic methods, working from within a particular community, rather than looking on from the outside (Latour and Woolgar 1986, Berkenkotter and Huckin 1995). During the study, different sets of data were analysed:

1. *Written texts*, comprising engineering proposals, numerous reports, technical notes, log book entries, and other technical documents. Texts (and genres) of particular interest to design engineers are examined in later chapters.
2. *Spoken data*, comprising 25 recorded interviews. During these inter- views and other conversations, engineers shared candid views about their work and use of language in various situations.
3. *Research journal entries* written over the course of the study. These are included throughout the book to describe the engineering workplace from a more personal perspective. The extracts tell stories, providing insights into the events surrounding the texts and the engineers who produce them.
4. *Email responses from 59 engineers* who participated in an email survey. A variety of engineers participated in the survey, which was volun- tary and primarily devised to investigate the writing tasks they have to perform at work. Survey findings are discussed in this and later chapters.

The engineers who provided information for this book perform a myriad of complex tasks as part of their jobs, including, for example, designing solid-state silicon 'gyroscopes' for use in cars; choosing the best glue to use, including testing the stickability and reliability of various adhesives for use in the stratosphere (to glue a component to a metal casing on a rocket); and writing software code to program screen displays for pilots in aircraft. The engineers work on both hardware and software systems, with the majority being practical engineers, concerned with the design, production, and maintenance of a product. There is a smaller number, concerned with theoretical modelling and research, sometimes referred to as 'boffins' and looked on with respect for their high-level knowledge of physics and mathematics. A few are primarily concerned with management and commercial aspects. Finally, a small fraction are women. Of those who took part in the email survey, with the exception of five, all are men.

1.2 Why they became engineers

As part of the email survey, engineers were asked why they chose to join the profession. Their explanations reveal that a pragmatic outlook and desire to be technologically creative was the main motivation, the vast majority having made a deliberate choice to become engineers. Of the 59 who took part in the survey, only 3 had either drifted into the job or chose it for financial reasons. A significant proportion of engineers had a strong sense of engineering vocation when young, making a choice of career at school. The majority (95 per cent) became engineers because they liked science and mathematics, had a keen interest in how things work, and wished to work in a job that was practical and involved problem solving (Sales 2002: 1–30). Nearly 30 per cent of all engineers described themselves as not having been particularly inclined towards the sciences, claiming to have more balanced inclinations: they are either neutral or positive in their attitudes towards English and Arts subjects, but have an overriding liking for, and, in some cases, a love of, science and mathematics:

> I very much enjoyed English in school and the science subjects too. I was pretty good at all my subjects but it was much more fun to play rugby and experiment with electronics after school rather than to try and read Shakespeare plays.

> I chose engineering mainly because I like problem-solving. I did not dislike Arts subjects, but found them to be not as challenging as scientific subjects.

One respondent explained he had been equally good at Arts and Science subjects at school, but had always wanted to join the Navy:

> It was a natural progression. I'd wanted to join the Navy when I was a boy. I'd always liked fiddling with machinery and the Navy let me do this.

Many engineers have this interest in (or, in a many cases, a passion for) tinkering with machinery and making things. They tend to feel an affinity with things technological, a few suggesting that engineering is 'in the genes':

> I became an engineer because I was always curious about 'how things worked'. I did enjoy Maths more than English at school because I found English boring, therefore I didn't work as hard at English (and other Art subjects) as I did at Maths and the Sciences. I did not make a conscious decision to become an engineer because I didn't like English.

> I enjoyed science (especially physics at school) and saw engineering as a useful example of applied science.

> Very interested in science – I see engineering as one of the best ways to exploit this.

If the email responses so far strike you as having been written mainly by men, you would be correct. When the survey was conducted, out of a total workforce comprising around 350 engineers, only 10 of the engineers were women. These numbers reflect the lamentably small proportion of females in engineering, with about 5 per cent of professional British engineers being women, and 10 per cent in the United States of America. All 10 were contacted, since there were so few of them, and 5 of the 10 responded. One, reflecting a generally held view that women are better at language-related skills, suggested that female engineers were probably equally good at Arts and Science subjects, and, furthermore, were probably more proficient writers. However, small though the female sample is, this belief is certainly not substantiated by the findings of this survey, as evidenced by the following responses, all provided by female engineers:

> I did prefer maths, physics and art at school as opposed to subjects involving lots of writing, e.g. Biology, History and literature.

> Performed better in Maths and science subjects at school, and less well at writing and spelling.

> This may be more of a female trait – I enjoyed both maths/physics and English language/literature, but it was predominantly my love of science that swayed me. I also enjoyed practical subjects and wanted to escape from an office-based career.

Around 30 per cent of respondents had performed better in Science (and, in several cases, Mathematics) and less well in English at school. All the engineers in this category clearly believe themselves to be poor (or underperformers) in Arts subjects, and that a scientific inclination was a deciding factor in their choice of engineering as a career, as the following responses reveal:

> a) I became an engineer because it was the easiest way to express my creativity. b) I fit into the Maths stream c) I was not good at languages at school. The mechanics of both reading and writing never came naturally to me, consequently, it is much easier for me to do engineering type jobs which involve a lot less of this.

> I am a Maths stream person. I was good at Maths, Science and practical subjects at school. I always found spelling difficult and never performed well in English.

These comments lend some support to a kind of engineering lore that exists, perpetuated by the engineers themselves, about how badly engineers communicate. A variety of horror stories circulate in the engineering community about how their inadequate writing skills lose the industry millions of pounds each year. For example, it is believed that a significant proportion of the estimated annual wastage of £2 billion (of an annual defence budget of c.£20 billion in the United Kingdom) is attributable to the poor writing of specifications and requirements, and badly written contracts (Kincaid 1997: 54). When asked about the accuracy of these stories, a senior engineer responded with these words, expressing a view that is fairly typical:

> Oh yes, absolutely – they don't call it a problem – this is an issue, an issue [sharp intake of breath] and people say they're engineers, they're not going to write documentation unless you stand on their neck. And when they do write it, it's a moveable feast. It might be good, it depends on the person. I mean you get people – you've seen Rick's stuff – who say, 10 o'clock I went to the toilet . . . and he goes on writing, and records the last 15 years of his life on that project. I've known half a dozen people like that . . . (Author's interview data)

Clearly, such stories are a misrepresentation of the full picture, but they are told and retold all the same. The fact that nearly one-third of the engineer respondents believed they performed less well (or were low achievers) in English would seem to lend credence to the view held by some lecturers in engineering that their undergraduates are deficient in English language (EL) skills, and, because of this, have chosen to follow a science and mathematics route as an avoidance strategy, in the mistaken belief that they would not have to write very much. The fact is, though, that they do have to write, they have to write a lot, and what they have to write is important.

1.3   Engineers have to write a lot

Let us now turn our attention to temporal matters. It is a little-known fact that writing is a time-consuming activity for many engineers. Nearly 50 years ago, Hicks commented on the sheer volume of paper and text that was produced in engineering offices:

> The output varies from a single-page maintenance instruction to a volume of five hundred or more pages covering an important scientific or engineering project. Operating maintenance and instruction manuals for some advanced missile systems run to several thousand pages, weigh 100 or more pounds, stand 5 feet high, and cost almost $1 million to prepare. (1961: 2)

When Hicks wrote this, much of the burden of writing was borne by technical authors working with pools of typists, professional draftsmen, illustrators, and printers. He claims that the 'normal duties for which the engineer or scientist is employed are not writing' (ibid.: 3), unlike the findings of this study which reveal that engineers can spend at least 50 per cent of their working time on writing. Now, in the modern engineering office, with working roles less clearly delineated, engineers play a much larger role in the writing process, sometimes taking responsibility for the production of a whole document, if not large sections of it.

Over the last two decades, engineers' work has changed dramatically, concomitant with developments in office technology. Their preoccupation with documentation, and the importance of writing about the product, continues unabated, not having changed much in this time. However, what has clearly changed is the way that engineers have taken on more responsibility for writing, having been given their own personal computers. The typing services that used to be provided by secretaries and the 'typing pool' have disappeared for those who are not in senior management positions.

These days writing plays an important role in engineers' work, to the extent that it can be extremely time-consuming. However, until the email survey was conducted, there was a lack of information about actual amounts of time spent on writing. Part of the survey involved discovering the extent of the writing done by engineers, who were asked to provide estimates in percentage terms. They were also asked questions about the types of writing tasks being performed and documents produced (these being the focus of later chapters). The main reason for asking these questions was to pinpoint those texts and documents that engineers find time-consuming and to gauge whether or not an investigation into written communication was justified. In the event, results showed that writing takes up a significant amount of an engineer's time and that certain documents are problematic.

The results are impressive or stark, depending on your viewpoint: slightly more than half of the engineer respondents, 50.1 per cent to be precise, spend between 30–60 per cent, and a further 15 per cent of the engineers spend more than 60 per cent of their time on writing at work. Figure 1.1 provides an overview of amounts of time spent on writing,

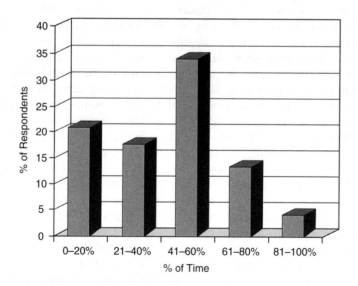

Figure 1.1   Time engineers spend on writing

providing a breakdown in 20 per cent time-bands of the figures for every respondent to the questionnaire.

The picture emerging here contradicts the stereotypical image of engineers engrossed in tinkering with machines and making things. However, it coincides with findings of another study carried out on scientists at the Salk Institute. The study, which concerned biologists rather than engineers, showed the major output of the scientists' work to be documentation rather than scientific experiments, and writing scientific papers to be central to their working lives (Latour and Woolgar 1986). The figures lend support to an idea that develops in Chapter 4 about the importance attributed to certain documentation, and the emerging realisation that text is, in fact, a substitute for the product, in that the product is negotiated and shaped through the documentation.

1.4   The customer

The customer looms large in engineers' work and holds a position of power and control over them. Rarely is the customer talked about except in impersonal terms, in much the same way as one might talk about 'the government', and, in fact, government departments in different countries may indeed be the customer. The customer, usually referred to in the singular and as a proper noun (and therefore capitalised in writing), is usually the company or person who commissions (or buys) goods or services designed and produced by the engineers. The customer is rarely referred to by pronouns, such as 'he', 'they', or 'it', except when the engineers are talking about, and know, a person who represents the customer. Put simply, in engineers' minds, the customer is the paymaster, or is a group entity comprising emissaries of the paymaster. Engineers are ever mindful of the customer's wants (and needs) and strive to cater for them as satisfactorily as possible, for, in the end, their livelihoods depend on the customer being satisfied. This probably accounts for exhortations by managers from time to time that they should be less preoccupied with the product and become more customer-focused. However, engineers' informal talk in day-to-day discussions reveals otherwise: in their working inclinations, they are essentially product-centric, that is, they are naturally pre-disposed to thinking about the product. In the day-to-day tasks that they perform, the messages they compose, and their unguarded discussions, engineers are engrossed in the minutiae of the design of their product, how it can be assembled or compiled, and tested before delivering to the customer. This is perfectly understandable, and it is far from the case that the customer is ill-served or lost sight of, although it has led to some prevarication over terminology later, where choices had to be made between using product-centred or customer-centred terminology: 'product' versus 'solution', 'product-support' versus 'customer-support', and 'selling-point' versus 'benefit', are some examples.

1.5 Types of engineer

Research journal entry: 'Good God No, I'm mechanical'

... I then ask Nick if he's talking about log books. He says he doesn't write logbooks. He can keep a record on his computer, and that's all he needs. He writes RESs instead, which, he says (straight faced but tongue-in-cheek) are what engineers call 'rough engineers' sketches'. The electronics boys, as he refers to them, keep logbooks because they have to keep track of so many (said with special emphasis and knowing look) changes. When I said: 'So you're not an electronics boy then?' he said: 'Good God, no! Give me something I can hit and knock about. I'm mechanical'. I now think of him as Good-God-No-Nick, because he says it so often and with that special rise-fall intonation. A lot about his reaction, I think, relates to the fact that he is a different type of engineer, ie 'hardware' as distinct from 'software'. Very different beasts. Engineers do tend to separate themselves into the different engineering disciplines, and see themselves as belonging to distinct groups.

[Author's comments: (1) 'RES' actually stands for Registered Engineering Sketch. (2) The significance of 'changes' is discussed in Chapter 5.]

1.5.1 Snap-shot of an engineering work-force

Engineers themselves provide the most realistic impression of their jobs. When asked to describe their work, they refer either to the type of work they are doing, to the official title of their positions in the company, or to their qualified status, as exemplified by the following responses:

'Software', 'Graduate Electronics Hons Engineer',
'As for my discipline, I'm a mechanical engineer'.

'Quality Assurance with electronics background'
'Support. Control engineering originally'
'I am employed as a Systems Engineer but I am qualified as a Chartered Electrical Engineer'
'Technical Director (ex-Systems Engineering)'
'By formal training an electronics engineer. By career, a systems engineer.'

Some engineers refer to the fact that their work involves different areas:

'I am a hardware engineer – although I get to have a go at systems, software, mechanical – whatever needs doing!'
'Labelled as Systems Engineer . . . reality is sitting on the fence between AR [Applied Research] and Business.'

By far the largest group in the survey comprises electronics engineers (25 per cent), four of whom describe themselves as being concerned with design, systems design, or design and testing. The second- and third-largest groups are software and systems engineers (20 and 18.3 per cent, respectively), whose work is inter-dependent. In very simple terms, systems engineers deal with design and drawing up requirements, and the software engineers with implementing those requirements (see Chapter 5). Systems engineers also write requirements for 'hardware' engineers to implement, for example electrical, electronic, and mechanical engineers. Mechanical engineers make up a smaller proportion of the respondents (8.3 per cent), with others (less than 5 per cent each) working in: production, support, mechanical design, manufacturing process, test equipment design, optical, control systems, and metallurgical and materials engineering.

For the purposes of this book, we can simplify the complex picture of the various types by narrowing them down to five major categories, described here by a support engineer in 'lay-speak':

Mechanical – 'designs the casing'
Hardware/Electronic – 'designs the circuits and innards'
Software – 'designs the software that makes it work'
Systems – 'integrates all the above, and makes sure the whole thing works'
Support – 'looks after the system, providing help and maintenance when it is being used.'

In the above list, it is possible to see two broad categories of engineer, which for the sake of simplicity and practicality are referred to as design and support engineers, that is, the first four categories and final category, respectively. Both design and support engineers contributed ideas to this study, bringing a different perspective and different views, although many more design engineers have been consulted than support engineers. This is because design engineers are the focus of this book. The role of support engineers is pivotal in post-design phases, when they become the mainstay and primary source of reference for the customer once the product starts to be used.

It would be more accurate, in fact, to place the systems engineer at the top of the list, since he develops the functional concept of the product, the detail of which is developed by the other engineering disciplines. However, this simplistic portrayal will suffice for the purposes of this book, which will examine particular communication tasks that engineers have to perform at work. These tasks, which most interest and preoccupy them, will emerge in later chapters.

Design engineers and support engineers

Figure 1.2 shows the symbiotic relationship that exists between three particular types of engineer: design, production, and support engineers. It depicts their roles to provide, at a glance, an understanding of how much time and effort they expend during the design and production of the product. It is compatible with various, more complex, diagrammatic representations of the product life-cycle, and maps onto the diagrams in Chapter 3, in particular Figure 3.4. Design engineers receive the most attention in this book, and it can be seen that they are primarily involved in the early stages of the product life-cycle, their work tapering off as the product develops and as support engineers assume more responsibility.

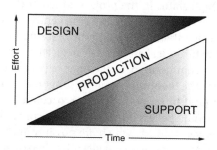

Figure 1.2 Work expended by different engineer categories

(The work of production engineers, however, is beyond the scope of this book, since they are concerned with the manufacture of the product, designing and testing the method and equipment used.)

Figure 1.2 is a fair representation from an engineering perspective, rather than from a financial, marketing and sales, production, or supplies/resourcing perspective, although these would throw up something very similar.

Design engineers have a tendency to be highly specialised and focused on arcane features of design. Of course, not all design engineers are narrowly focused like this, but as a general rule, this observation holds true. The very nature of their work, and their work practices, and procedures encourage design engineers in this tendency. Their high status is understandable, for it is the design engineers who are fundamental to winning new business for an engineering company. In any bidding war, where companies compete for business by submitting proposals to the customer, it is design engineers who must devise the most attractive design (or solution) with which to win the bid. Proposal writing is a central activity in many engineering companies, and it demands creativity from the engineers who concentrate on the design to the exclusion of anything else. (Engineering proposals are examined in later chapters.)

Parallels can be drawn with the building industry where the acknowledged 'king pin', a term sometimes used by members of the building team, is the architect. The architect is responsible for the overall design concept and converting it into a building or other structure. He has the responsibility for creating a blueprint of the design which others then use to construct the building. Behind the architect functions a team of consultants of various types – engineers, quantity surveyors, and building contractors – but it is the architect who has the kudos and the highest prestige within any building project. A similar situation exists within the engineering environment: the design engineers are the 'king pins', enjoying high status in the project. Much of the creative activity of the project team, including text creation, centres on catering to their needs and supporting them in devising the best solution. In the process, others' needs may be under-resourced or neglected, which is a situation support engineers commonly perceive themselves to be in.

If it were not for the involvement of the support engineers in the early stages of design (and in proposal writing), it could be argued that the design engineer would be less mindful of issues concerned with the use of the product. Several stories, which are possibly apocryphal, circulate about design blunders. One, which is gleefully cited by support engineers to show how the problems would have been averted had they been consulted, concerns an anti-tank weapon which was designed to be used in forests in cold regions. When it had been built and was being tested in the field, the user found he could not operate it with gloves on; nor could he see the controls in the dark. Although all engineers tell stories of this type, support engineers are particularly good at telling them. It helps to compensate for their perception that they are undervalued and have lower status in the engineering workplace, and also their belief that they are consulted less than they should be in matters of design. The investigation into design and proposal documentation described in Chapter 9, which includes the cinderella domain of product support, would seem to confirm this impression.

Aspiring engineer writers and frustrated technical authors

In-company technical authors can play a pivotal role in preparing and writing formal company documents, for, to extend and borrow Swales' coining (1996: 194), they are 'textographers' with expertise in working with text. They co-ordinate the (sometimes numerous) contributors, collating the various (far from homogenous) textual contributions they receive, and compile documents that are professionally presented and read coherently (Austin 1990). As mentioned earlier, with the introduction of modern working practices, secretaries have almost disappeared and, instead, engineers have their own computers, providing word-processing facilities and other software. With new, more independent, methods of team working, a combined focus on both product and customer, having to respond to customer queries themselves (rather than a secretary doing it), and the need to do their own typing/word-processing, it is hardly surprising that the engineers should produce their own drafts of text, or take the initiative in producing documentation. Rather than produce a sketch outline for the technical author to interpret and flesh-out, as they used to do, it is now more convenient for them to commit their thoughts to the screen and compose their own drafts. All this has had an impact on technical authors, who are used to taking responsibility for document production. They find they are called on less frequently to draft documents from scratch, which is what they prefer to do, and where they see their expertise lies. Instead, they now deal more often with cosmetic aspects of writing: formatting text, integrating graphics, 'tidying-up' sentences, and generally checking grammar and spelling, and editing what engineers have written. Some see this as a retrograde step and feel, as the editor of the 'Communicator' (a journal for technical authors) puts it, 'dismay at the potential effect on the role of technical author' (Newell 2005). Being presented with texts that engineers have written themselves more often these days, authors believe, limits their professional practice and restricts opportunities to use higher-level writing skills. It also, in their view, erodes their status in the eyes of some of their colleagues who fail to understand the true role of the technical author.

Technical authors have expressed the opinion that design engineers cannot write as well as they design, and that by writing their own technical descriptions and other kinds of writing, they are using time that would be better spent on work they have been trained, and paid, to do. The authors feel that much of the time and effort they spend on rewriting and editing engineers' compositions is a poor use of their time, and ultimately affects the quality of the finished document. They are sometimes exasperated with the texts they are given to work with, finding them illogically structured, often grammatically incorrect, and stylistically inelegant. By the very nature of their jobs, technical authors offer engineers a documentation and writing service, and, unfortunately for them, usually find themselves having to be reactive rather than proactive, too often responding to requests, rather than being involved at the outset of document production.

A tension clearly exists over authorship and ownership of text, with the engineers who are concerned with design also concerned about any text that is produced about their design for other people to read. They are proprietary about any text relating to their product intended for an external audience. In this book it is suggested that text performs the role of substitute for the product in design documentation (see Chapters 3 and 7), where the text is treated as being the product for certain intents and purposes until the actual product is produced. It is therefore understandable that engineers should feel responsible for, and have a 'mother–hen' attitude towards, any text that describes their product.

1.6   Creativity versus restriction

Engineers attribute different values to different texts, holding particular documents in esteem and disregarding, or even disparaging, others. Figure 1.3 represents the views of an engineer, who values any documents relating to the engineering development of a product, but who dismisses as bland and restrictive most others.

His is a design-focused view of text, showing his interest in the creative aspects of product development. Not all engineers would agree with his description of 'official' documents as being bland, however, although they would accept they are restrictive. They tread a fine line between

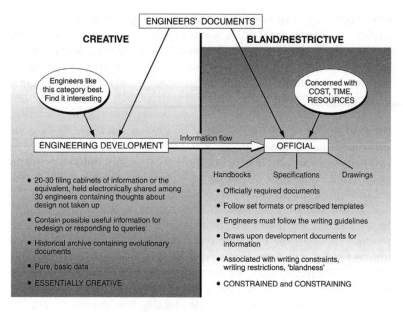

Figure 1.3 Engineer's view of documentation he has to produce

being creative and being controlled by regulations and procedures that are part and parcel of any modern 'high-tech' workplace. Subsequent chapters explore how engineers deal with this tension, exploring the issues that arise, their communication tasks, and the texts they produce when working on key documents at work.

Engineers are solution-oriented

When asked about their writing activities at work in the survey, a few engineers listed, among other things, writing poetry and post-it messages. Post-it messages may appear to have some relevance to work, but composing poetry has less obvious work connotations. However, one of the engineers explained why he liked to write poetry:

> Sometimes it helps to spend a few moments on something creative, which has got nothing to do with the problem you are working on – helps clear the mind.

It would be misleading to perpetuate stereotypes and myths that circulate in non-scientific circles. Arts students and lecturers, for example, have been known to refer disparagingly to engineers and engineering students as 'inarticulate nerds', or some such uncomplimentary epithet. By the same token, engineers are vexed at times by the scientific ignorance of those from non-scientific backgrounds, and view disparagingly their want of self-discipline, for, and let there be no doubt about it, one of the distinguishing features of engineers is their self-discipline and circumspect behaviour. Their minds and bodies they regard, not unlike the things they design and make, to be sorts of machines that can be controlled, modified, and maintained. They regard thinking processes as controllable, and capable of being managed to obtain optimal performance. Thus, composing poetry or music for a short while at work will aid concentration: all the better to achieve work goals.

It would therefore be true to say that by training, and possibly by inclination, engineers are solution-oriented, in that they strive to produce the item the customer has ordered. In a sense, they are not unlike tailors, especially in the case of 'bespoke' products, like, for example, a navigational tool for a car or a ship. Having agreed on the details of the tool, engineers engage with the task of designing and making it, despite the many obstacles that they usually encounter. In pursuing their goals, some may be considered almost obsessive, because of their preoccupation with the minutiae of design. Such rather contentious observations have been made by those outside engineering. Engineers have explained that their absorption is understandable, since solving mathematical problems or difficulties with software coding can monopolise their thoughts, even in non-work social and domestic situations, making them appear distant or socially gauche. Putting such considerations of behavioural traits aside, however, it is an irrefutable fact that engineers are goal-oriented, perpetually solving problems in the search for solutions. 'Solution' is the word so often on their lips.

Sensitivity to language: 'solution' versus 'problem'

Design engineers, and their managers, do not use the word 'problem' very much in their writing and everyday work talk, some acknowledging that they consciously avoid using it. These are engineers in direct communication with the customer, who needs to be persuaded and reassured about the effectiveness of the product they have designed. They explain that 'problem' has negative connotations that sit uneasily with the sort of mindset and positive attitude needed to perform the task. So focused are these engineers on working towards a solution, that 'problem' tends to be used only when the obstacle seems insurmountable. However, in other engineering domains, for example, materials, manufacturing, and production, engineers use the word more frequently, simply because they interact with colleagues, and not the customer, in problem-solving tasks. It is clear the different audience and context calls for a different linguistic register.

'Solution' ≈ 'product'

By contrast, the positively coloured 'solution' is more popular with engineers, especially with those who are in direct contact with the customer. They use 'solution' in speech and writing to refer not only to the answer to a problem, but also to the product itself. In the case of proposal writing, for example, or talking about proposals, engineers prefer 'solution' to 'product' for referring to what they are trying to sell to the customer, because it portrays the speaker as having a customer-oriented approach, that is, as one engineer put it: 'By using "solution", we show the customer we are focused on coming up with the best solution to his needs'. Again, as with 'problem', it is clear that context of use and audience determines stylistic choice of words. It is possible to see the attraction of 'solution' as a cover-all term, useful for engineers to convey to the customer that they have the answer to his problems, and that they are working in the customer's best interests. 'Product' conveys the sense of tangible objects, whereas 'solution' is more abstract and intangible, with much wider (non-engineering) applicability. In discussions about the complexities of engineering design, it is sometimes more appropriate to use 'solution' when a generic sense is intended, or where the speaker is referring to the whole answer to a customer's requirement (or need), which may include different aspects, for example, a maintenance plan, or training for users. However, tellingly, 'product' is the term engineers use amongst themselves in meetings and informal discussion at work, although they tend to use 'solution' more often in dealings with the customer.

Subjectivity versus objectivity

Those teaching in the Arts see as their role the fostering of a person's ability to express personal opinions and to give vent to self- and emotional expression through some kind of 'creative' outlet like music, painting, or literary composition. It could be said that such artistic expression requires egoism to a greater or lesser degree, depending on the artistic work, in which centrality of the 'self' is a natural starting point. The results of these expressions may be manifest as individual or group efforts, as in the case of a painting, poem, or a novel, say, or an orchestral performance or dramatisation of a play script. In the sciences and applied sciences, such self-expression is generally anathema. Engineers baulk at the slightest hint of 'subjectivity' in their language use at work, and can be strident in their criticism of it. As a body, they resist strongly (and, some in the Arts and Humanities believe, stubbornly) any hint of personal opinion, consciously avoiding any use of the personal pronouns 'I', 'you', 'he', 'she', and 'we', for, as one engineer put it: 'the pronouns get in the way of objective thought'. They prefer to use passive rather than active verb constructions; for example, they prefer '"Gyroscope" is used generically...' to 'We use "gyroscope" generically...'. Kirkman, however, advocates a flexible approach, advising engineers to use a more personal style with certain documents, like reports or papers (1992: 73), but such advice tends to fall on deaf ears so far as design engineers are concerned. Van Nostrand similarly observes that in corporate documents in the American defense sector, and in corporate documents generally, the first-person singular personal pronoun ('I') is noticeably absent and 'the texts share a deep formal resemblance' (1997: 137). Dobrin refers to this stylistic convention in his discussion of objective writing, commenting that he finds 'the whole stipulation of formal objectivity puzzling', because the use of particular so-called impersonal, objective language 'doesn't confer objectivity, though it can be a shield' (1989: 35).

Engineers hold particularly firm views about the need to write clearly and objectively. Essentially, this demands that the writer should quell any egoistic tendency in writing, or at least should mask any such tendency. As the engineer cited earlier explained:

> Engineers are not impressed by anything that isn't a fact. It's difficult to be totally objective if you're using active verbs and personal pronouns. In our writing we go for the suppression of self, not the expression of self.

This is easier said than done, however, as there are times when engineers are caught between two stools. These are times when they need to reconcile their aim to be clear and objective with the obvious need to be persuasive. For example, when writing engineering proposals, they find themselves in the curious (and uncomfortable) situation of having to be persuasive without being too obvious about it. The rather slippery notion of persuasion and the stylistic and cultural conflicts facing engineers when working on proposals are discussed in later chapters.

1.7 Communication skills: engineers' aspirations

In one-to-one interviews, engineers have discussed stylistic aspects, expressing a yearning to be more articulate and elegant in their use of language. A few describe how they notice others' language performance, particularly in meetings (Sales 2002: 2–7). They mention by name articulate colleagues whose writing they admire, or whom they have noticed hold the floor in meetings. They say they wish they could hold people's attention in a similar way, regretting their lack of eloquence or conciseness. As one interviewee put it:

> I think what I'm maybe not so good at doing as I ought to be is saying things concisely, saying things in fewer rather than many words and I often find when I'm talking to people I can sense sometimes . . . a . . . you're saying too much, you're going too far, and yet I can see other people that can command attention by saying fewer words because what they say is straight to the point. I think that would be a nice thing to work on, to be able to say things concisely and accurately. Get your meaning across in fewer words.

Rather larger numbers of engineers say they find writing stressful, not being confident in their writing abilities, and being concerned about grammar and vocabulary. Some are self-critical, denigrating their spoken and written expression by describing themselves as lacking in conciseness, being mediocre, and not having the 'right' words. In the case of writing, apart from mentioning the problems they have with grammar, spelling, punctuation, and the like, they also worry about having mental blocks and spending too much time composing relatively short pieces of text. Professional writers would claim that this is all a natural part of the writing process. Other engineers talk about resorting to copying from other documents as a strategy to help them complete a writing task, one referring to it as plagiarising and describing it as an exercise in damage limitation. Associating such copying with plagiarising and feeling a sense of guilt are needless and unnecessary, however, for this is a common practice throughout the commercial sector when drafting documents, especially certain commercial letters and reports (Marshall 1986: 136).

A few others, who work on commercial and administrative aspects, talk about the 'drabness' of their writing, finding it difficult to write in a friendly or positive way without being 'smarmy', and difficulties in composing a rejection letter without making the reader feel slighted, and without appearing rude or indifferent. One engineer laments his over-formal style when writing letters and memos. His problem, he believes, is appearing to his audience as a distant and unfriendly stranger, even when he is, in fact, writing to people he knows well and with whom he has been communicating over several years. He says he does not know how to be friendly in a letter. Judging by his writing, it appears that he is indeed in the habit of expressing himself formally, as the following extract from a memo shows:

RE: *MAJOR BID PERFORMANCE WORKSHOP*

As discussed at the end of our workshop, please find attached tran-scriptions of our charts, and my draft memo to TCB/DMD/IDHS/PC which encapsulates the way ahead.

Please provide me with any comments on the attached by COB 6 June; in particular any disagreements or further suggestions.

[COB – close of business]

The piece exudes formality through: clauses beginning with 'as discussed', 'please find attached'; the use of (lengthy) initials and acronyms, 'TCB/DMD/IDHS/PC' and 'COB'; a depersonalised style conveyed through omission of any reference, by name or use of pronouns like 'you', to those being written to; words like 'discussed' (instead of the prepositional verb 'talked about') and 'provide . . . with' (instead of the more colloquial 'let . . . have' or 'give'); and the use of punctuation (semi-colon) and complex sentence structures usually asso-ciated with formal writing. As it stands and considering that close colleagues were the target audience, it certainly comes across as formal. After examining his writing together, the engineer and I agreed he had developed the formal style during years of writing engineering docu-ments, when his work involved more engineering and less administra-tion. As we shall see, engineers and, more particularly, design engineers are most concerned about expressing technical description in writing.

Another engineer finds writing so difficult that he likes to ask others to do it for him. However, as mentioned earlier, changes in work practices over recent years have thwarted this tactic: a gradual reduction of secret-arial assistance has led to the depletion of such writing support and seen the devolution of writing responsibilities to groups of engineers that are, in effect, writing teams. As a consequence, engineers themselves are now responsible for much of the written output in engineering companies.

There is ample anecdotal evidence of muttering in engineering circles about how badly engineers write. It is engineers themselves who are doing this muttering, and it is they who are critical about the writing produced by colleagues and peers. Anyone hearing such opinions would gain a general impression of negativity so far as engineers' writing is concerned, and certainly it seems to be quite accepted amongst engineers and in society at large that this is a fact. However, this blanket judgement is worthy of closer scrutiny, if only because, unquestionably, there exists great dissatisfaction (without exaggeration) about documentation across the industry, with complaints about how expensive, time-consuming, and costly it is when mistakes are made. But can engineers' inadequate writing skills really be blamed for this malaise? As this chapter has shown, only a quarter of engineers seem to have difficulties with writing, and so it is more likely that the answer will be found through examining communication tasks, and working and writing practices in the workplace. This is one of the main purposes of this book, and as later chapters reveal, the story is a complex one. It is possible that, by denigrating the writing done by others of their profession, engineers are perpetuating a possible myth.

1.8   Final comment on the discourse community and aims of this book

It would seem not much has changed since C.P. Snow's 'Two Cultures'. There still exists today, as there did when Snow gave his seminal lecture in 1959 on the cultural divide (some would say, chasm) between the Arts and the Sciences, a broad misunderstanding between the two camps (Snow 1964). Some see this in stronger terms, as amounting to a two-way antipathy between those working in the Arts/Humanities and those in the Sciences. As an applied linguist, I have had a foot in both camps over a decade or more, and through this book wish to achieve two main goals. First, to describe to aspiring engineers (and the customers who judge what they say and write) the complexities of the communication tasks they will have to perform as working engineers. Secondly, I would like to bring a better understanding of engineers and their texts to other applied linguists who are interested in analysing texts in the engineering domain.

Overall, a striking inference to be drawn from engineers' responses to the survey is that they see themselves as belonging to a distinct group in society, with cognate qualifications, similar interests and ideals, and common goals. They may be regarded as forming a distinctive working community. Swales (1996: 20) refers to different terms we could use to refer to such a group, discussing variants of 'discourse community' (which he refers to as a troubled concept) like, for example, 'rhetorical community', 'disciplinary community', and the more recent 'community of practice'. He states that such communities are essentially occupational or recreational groups that are 'somewhat different' from the sociolinguistic 'speech communities' which have their basis in geographical location or delineation (ibid.), and that 'In effect, in discourse communities, communalities reside in what people do rather than in who they are.' In spite of his apparent ambivalence towards the term, Swales' description of a discourse community is clear and useful:

> Discourse communities are sociorhetorical networks that form in order to work towards sets of common goals. One of the characteristics that established members of these discourse communities possess is familiarity with the particular genres that are used in the communicative furtherance of those sets of goals. In consequence, genres are the properties of discourse communities; that is to say, genres belong to discourse communities, not to individuals, other kinds of grouping or to wider speech communities. (Swales 1990: 9)

It is important in a study of professional communication like this, be it from an applied linguistic or engineering perspective, to work within a recognisable discourse community or community of practice. At the heart of any study like this must be the oral and written tracts of language that are produced, manifest as texts or textual genres. It is these texts and the stories that surround them that form the basis of this book. Engineers are the main players in the stories, and their views have enabled an ethnographic approach to be taken to this study of their communication tasks at work.

They have revealed aspects of their work that merit particular attention, and have brought to my attention certain texts (spoken and written), documents, and other verbal outputs considered important in their discourse community. In the following chapters I examine those texts and documents that preoccupy engineers most, and tell their side of the story. There has been a long tradition within applied linguistics, particularly English for Specific Purposes, of research concentrating on a cluster of genres and text-types, namely, reports, academic journal articles, business correspondence, and academic assignments. Engineers have little interest in any of these. Therefore, this book also pays them scant heed.

2
Engineering Practices and Procedures

2.1 Engineering procedures

2.1.1 The burden of responsibility

It could be said that, through our efforts at work, we are contributing to the wealth and well-being of society. However, not all our contributions have direct life-or-death consequences as is the case in professions like medicine or engineering. A flawed design could kill people: this is ultimate burden of responsibility of the engineer. So their creations must stand the scientific scrutiny of proofs and tests if they are to be used by other humans. It is this burden of responsibility that has contributed to the importance of procedures in the working lives of engineers.

As a general rule, if asked a question about writing specifications, engineers' automatic reaction is to consult company procedures. In fact, any specific question about writing or documentation has this effect, because engineers rely on procedures to guide them in their work. There is general expectation among engineers that procedures ought to account for most eventualities that may occur in their everyday work. It follows, then, that when a new work activity is envisaged, those supervising the implementation of the activity will spend much time drafting a procedural framework for it. The aim of developing a framework would be to ensure standardised working practice and maintenance of quality, with the underlying expectation that the engineers would operate within it.

On an abstract and theoretical level, then, there is an unquestioning acceptance of the need for prescription so far as working practices are concerned, and the belief that if everyone follows procedures, life would be easier, neater, and tidier for everyone. To a non-engineer applied linguist like me, it is a striking feature of this discourse community that engineers try to impose form on their own behaviour and activities at work, in much the same way as they do on the products they design.

2.1.2 Engineering methodologies: the importance of procedures

An engineer gave the following explanation about the importance of following procedures:

> It's important to know what you're doing, questioning whatever you do, and testing it at every step... people [engineers] say they do it anyway, but some of them fail to see the relationships between the different stages and it's all too easy for something to be missed out which may completely stymie the whole process.

He was talking about design procedures, although his words would be echoed across all engineering domains. The importance of the scientific approach and validation of the design through rigorous testing and proofs is fundamental to their work. It is this that provides the impetus for the continual revisiting of the procedures, which themselves come under scrutiny and revision in the perpetual cycle of refreshment and renewal of engineers' work practices. This belief in procedures is rooted in the fear that major problems may arise out of an inaccurate or incomplete design, which could lead to an inability to fulfil the terms of a contract. This, in turn, would lead to a dissatisfied customer and financial losses running into possibly millions of pounds. For this reason, companies with research and development facilities allocate resources so that staff time may be dedicated to pre-empting such problems, or trying to identify the causes of these problems.

More often than not, in attempting to solve them, they change the procedures. Procedures writing is a way of life for many engineers. They consider it unremarkable that colleagues may devote nine months, more or less exclusively, to a project on work methodologies, the outcome of which would be the design of a master plan for the way they should work. The rationale underpinning these projects reflects engineers' desire to produce an abstract specification of a process or processes, which would in turn give rise to different procedures, depending on the particular tasks needing to be done. Such work plans usually result in some kind of procedural document called, for example, 'Engineering Methodologies'. Those outside engineering often fail to realise the importance of methodologies and procedures, or the extent to which planning them and writing about them plays an integral part of an engineer's work activity. With ever-changing technologies and the need to keep abreast of developments, engineers are caught up in a perpetual cycle of either writing procedures or following them.

It can be seen, then, that assumptions about the necessity of providing a procedural framework are fundamental, and documentation for this abounds across the industry. There is also plentiful and costly evidence for an assumption commonly held by engineers that there is a 'correct' way of doing something, with just a hint of distrust: there seems to exist a suspicion that fellow engineers may forget an essential 'step' or be slipshod in their practices, and that this must be catered for or pre-empted. There is also a concomitant assumption that it is both desirable and possible to account for and delineate anything concerning their work, to a degree of delicacy that I have not observed in other commercial organisations. Moreover, this quest of engineers to discover the 'correct way' encompasses, of course, the writing they have to produce. And there's the rub: engineers' writing has proved notoriously difficult to control, as later chapters reveal.

Unforeseen costs and losses have proved a strong motivation for organisations to find solutions, which, it is believed, mainly lie in changing work practices and improving the writing of specifications and requirement. Most engineering companies share the concern to improve in this area, as revealed by investment projects dedicated to improving the quality and reliability of design (and writing for design) procedures. An example of such a project is the long-term research programme at the Dependable Computing Systems Centre (DCSC), now based at the University of York in the United Kingdom. It is dedicated to eliminating flaws in the early design stages of computer systems. Publicity about the centre includes reasons for its establishment, one of which is couched in the following terms:

> The research conducted at the DCSC is intended to reduce the cost of producing these systems and reduce the development lead-in time for systems for which there is an ever increasing customer and market integrity expectation. (DCSC 1996)

The reference to 'customer and market integrity expectation' is a reference to the fact that both customers (those who commission the design and development of products) and users (who may be ordinary members of the public or the armed forces who operate and use the products) have ever-increasing expectations about health and safety standards, no matter how complex or intrinsically dangerous the product might be. Be it a component for a wheel chair, a naval weapon control system, or a fighter jet, society today expects the product to be designed and built to a high level of safety. The DCSC also refers to the need to reduce costs and development times, since these have proved a major cost bugbear to companies particularly on early projects where higher levels of safety were required. In the defence market, for example, which is currently experiencing greater competition, companies have to adapt to radically different tendering procedures, and even more demanding customers, who are no longer prepared to wait 5 or 10 years for a crucial piece of equipment to be developed. A major fear companies have is that they may become embroiled in projects which overrun and incur extra unexpected costs. Anything which is unplanned or not anticipated is anathema to the engineer, who yearns (usually in vain) for predictable outcomes.

The search for improved procedures has also spawned the development of software packages that are complex databases and information retrieval systems. To name three examples, there is Requirements Traceability Management (RTM), Dynamic Object-Oriented Requirements System (DOORS), and Lightweight Object Repository (Lore) for engineers to use in systems design. These programs are doubtless a consequence of engineers' distrust of 'natural English' (Chapter 5) and the disappointing lack of progress that has been made in the field of Natural Language Processing (Sinclair 2004: 192, and 2001). The programs are touted as helping engineers not only with the design and navigating the structure of their work, but also with the writing of specifications and test procedures. An interesting feature of these programs is their intention to control engineers' work behaviour. RTM is a case in point: it is essentially a management tool, which is used to keep track of changes that are made to a design, ensure that the necessary related changes are actually done (because there is always a knock-on effect of making even the smallest change), and ultimately ensure consistency of working practice. Any sacrifice to creativity the engineers have to make in order to control, or even constrain, their work behaviour is seen as a small price to pay for ensuring this consistency.

Engineers' evaluations of such programs are mixed, however, and claims about their efficacy regarded as dubious, but this does not dampen their enthusiasm in the seemingly never-ending search for the holy grail of control over design and specifications.

Research journal entries: Proving the job has been done properly

Two separate entries:

First entry: It seems engineers who keep logbooks do so to record the work they do when they are solving problems. Paul's just been talking about having to try out as many as 23 different resistors and carrying out various tests on them until a solution is reached. They need to record all the tests they do and show, step-by-step, how they ultimately reach a solution.

Second entry: Mark offered to let me see some of his old logbooks, although he says he rarely writes an engineering logbook anymore, because he records notes and comments electronically [he is a software engineer after all]. However, he still records details of tests he's done from time to time on paper, and it seems to be a serious business indeed. He mentioned a particular test for which he had handwritten records. The test had been witnessed by another engineer. Mark wrote it up in a special record book, had this hand-written record of it verified by a senior engineer, and then he tore out the top copy carefully, filed it and locked it away.

He acknowledges he does this in case the customer asks for proof that the tests have been done.

2.1.3 Engineers' attempts to control writing behaviour

A typical working day for an engineer much depends on the type of company he or she works for. There still exist engineering companies, and large ones at that, which are run on more traditional, formal lines, have rigid hierarchical personnel structures, and house employees in offices that encourage them to focus more on the tasks sitting on their desktops and computer screens than on interaction with colleagues at other desks, let alone in other offices. However, more modern (and, some might say, enlightened) companies have flatter organisational structures and occupy large open-plan offices that encourage intermingling and interaction of all categories of employees. In such accommodation, engineers take part in a mix of spontaneous, unplanned activity, as well as that which is prescribed and pre-ordained by company procedures, for, notwithstanding the informality of the modern workplace, company procedures still underpin professional engineering activity.

So, on the one hand, there is plentiful and varied human interaction, including face-to-face chat, telephone conversations, responses to unexpected problems, and ad hoc meetings, much as one would expect to find in any busy open-plan office. On the other hand, the engineers understand that their activities are governed by certain expectations of how things ought to be done. They assume that for every task they have to perform, there is a particular way to perform it, a 'method', if you like. There is a tacit understanding, which is shared by all members of this discourse community, that these methods are expressed in the form of published procedures which are produced with the express purpose of guiding engineers through the various activities they have to carry out. It is an accepted part of the engineers' working ethos that for every task needing to be done, a procedure should exist to provide guidance on how it should be carried out, and that those responsible should try to anticipate procedures that will be needed. As a result, a company's published procedures may be exceedingly varied in nature, providing, for example, guidelines on working with lasers, writing a product specification, claiming overtime, using chauffeur-driven cars, and obtaining a 'Hot Work Permit'.

One of the memorable features of early conversations with engineers in the 1980s and 1990s was how each, on being asked a question, would automatically lead me to what they commonly referred to as 'the procedures'. In one particular company, the procedures comprised more than two thousand pages held in sets of four thick lever-arch files. I would be taken to the nearest set and watched as the engineer thumbed through to find the relevant section. At that time, more than thirty sets of these procedures existed, scattered about the site. The procedures still exist today, but no longer on paper. They have been revised several times, and are now held electronically. What struck me in those early years was that I had not encountered a single engineer who had actually used the procedures to help him with writing company documents.

Research journal entry: Managing change and 'ticking off' work

Another round of company restructuring on the way. It's being done by a specially formed 'Change Team' this time. They've really worked hard at their documentation and consulted with everyone, bar the window cleaners. It's all about the product development process, and what the team calls 'Phase Gates'. This is quite an evocative term. Engineers like to come up with these. For 'phase gates', say 'reviews'. They're similar to my 'thresholds' [see Chapter 3] but more complex in their structure, because they've incorporated into each 'gate' a description of recommended work activities, which are really procedures for staff to follow.

It is interesting to see the different approach the company has followed this time to implement change, which seems to be based on some kind of blend of philosophical and business principles. They've spent ages developing a justification for the changes, and in true engineering fashion, a small team of engineers have spent more than a year consulting with managers and staff, analysing work practices, holding numerous staff forums, and devising various possible models of working practice and organisation. They have finally settled for the 'value-chain' model. This means they all have to focus much more on the customer, and by this they mean that engineers need to communicate more with the customer and that there are specific communication tasks they need to perform during any project. So more procedures for them to follow.

They seem to be using the term 'Phase Gates' as an abstract concept, not unlike some kind of metaphorical hoop the engineers have to jump through in order to progress to the next phase. These Phase Gates are intended to prepare the way for major changes to take place in work practices which will see teams of engineers broken up and reformed into 'value chains', a new name for what are in reality linear-like teams. They were in product-focused teams before. To my way of thinking, 'gate' is a bit of a misnomer, implying as it does a simple crossing from one phase to the next. However, in this company, 'gate-keepers' (teams of internal auditors) check that certain procedures have been followed and that certain tasks have been achieved, before the engineers can 'progress' so to speak. So, passing through the 'gate' could prove to be a fairly lengthy process in itself, a bit like trying to cross over a cattle-grid in stilettos. This checking, or rather, ticking-off against a list of job items, is really significant, because it seems inherent to an engineers' work at various stages. The ticking-off, or signing-off, of tasks influences the approach engineers take to their work, and, importantly, influences the way they structure their technical documents.

2.2 Writing procedures and guidelines

2.2.1 Written, and then often ignored

The writing done by engineers receives the same treatment as any other task they have to perform, with a plethora of procedures to help guide them, guidelines which aim to cover all aspects of writing, ranging from the way a document should be formatted through to the length and style of sentences. As is mentioned later in this section, procedural guidelines attempt to impose particular syntactic structures and restrict the lexicon from which writers may select words to compose their sentences.

It is common throughout the sector that engineers take it for granted that procedures should exist to guide them with written communication. They believe it is both desirable to establish uniform writing procedures and possible to devise standardised formats and styles, if only someone would take the time to plan them and write them out. There is also an assumption that model texts, if only someone would devise them, would improve the documentation they produce. This attitude is exemplified in the following statements provided by two engineers:

> I come from an engineering background and I like to work to a defined structure.
> It would be good if we could come up with a template for writing executive summaries. It would save us from searching all the time for the best way of doing things, and ensure consistency of writing style. We're always reinventing the wheel.

There is no doubt about engineers' reliance on procedures in their working lives; however, the situation surrounding writing is a curious (and difficult) one. In spite of their call for writing guides and templates, the fact remains that engineers can be resistant to what they see as being dictates imposed from above, interfering with the ways they like to write. A survey of the extent to which engineers refer to guidelines to help them when they are preparing engineering documents reveals that they seldom (or never) use the guidance available for writing (Sales 2002: 2–27). Only 6 engineers out of the 59 respondents (10 per cent) had ever referred to writing guidelines of any kind: 4 had used company procedures, policies, and instructions, and 2 had referred to British and American Military standards (American Department of Defense 1994, 1998, British Standard Institution 1998). This was not a surprising result, confirming the findings of an earlier study that identified the existence of several writing guides and procedures, but failed to find anyone who had actually referred to them. Examples of the guidelines unearthed in the earlier survey are Simplified English (a set of rules governing the writing done for readers who do not use English as their first language); Information Technology Security Evaluation Criteria (ITSEC); the company's own procedures, policies, and instructions; United Kingdom Ministry of Defence (MoD) Standards, commonly referred to as 'Def. Stans'; and American Department of Defense (DoD) Standards, usually called 'Mil.Specs.'

Engineers taking part in the survey were asked: 'I suspect you seldom (or never) use the guidance available for writing. Am I right?' Punctuation marks, especially the oft-used question mark and exclamation mark, were used to convey engineers' feelings, as revealed in the following verbatim responses:

> Yes, you are right!
> What guidance? – Does this answer your question?!!
> I didn't know we had guidance for writing!

A dearth of useful engineering-related reference material

Within the field of applied linguistics, there have been several significant studies of the roles of texts in organisations, and aspects of their discourse features (Odell and Goswami 1985, Paradis, Dobrin and Miller 1985, Myers 1990, Bazerman and Paradis 1991, Spilka 1993, Davies, Forey and Hyatt 1999, Bargiela-Chiappini and Nickerson 1999, among others), but there is a chronic shortage of relevant work in the different fields of engineering. Even within the general field of engineering, any (even remotely) useful work relates to civil engineering and the building industry, whereas mechanical, electrical, and electronic engineering appear to be neglected areas. The main libraries at three British 'red brick' universities with long-established engineering departments bear out this impression, yielding but two books on the writing of design documentation, both of which relate to the building industry.

This is the sorry situation, in spite of the fact that one of the universities is considered to be a leading university in engineering, an observation also acknowledged by one of its technical librarians. When asked if there were any books or reference material on engineering documentation and writing, he mentioned that the British Standards relating to writing about design had been removed from the main library to the store and would take some time to find. It was some small consolation to reflect that at least the engineer teachers and acolytes at the university manifest the same writing behaviour as engineers in the workplace: writing procedures are produced, but they pay little attention to them (apart from rare incidences). The reasons behind this state of affairs may lie with the fact that it is rare these days for British engineering undergraduates to be taught about the writing they will have to do at work. It was different in the 1960s and 1970s when particular writing tasks were an integral part of the engineering curriculum, for example writing log book entries and lab reports. It could be said that if it is not taught to engineering undergraduates now, it is logical no books on engineering writing skills are needed in university libraries. It must be assumed they will have to learn on the job (Winsor 1996: 19). The situation is very different in the United States of America, where there has been a long tradition of teaching writing at colleges and universities (Grabe and Kaplan 1996: 148).

2.2.2   Simplified English

It is possible that Simplified English is an example of writing procedures that have been ignored or disregarded in some quarters. Technical authors and other writers in companies belonging to the European Association of Aerospace Manufacturers (AECMA) are, in theory, bound by Simplified English procedures, which are, in effect, a set of formal writing guidelines. They were devised for manufacturers of civil aircraft, whose engineers and technical authors are supposed to follow Simplified English rules when preparing technical manuals.

Similar to any paper-based company procedures, Simplified English comes in a hefty ring-bound tome, comprising four hundred or more pages. Fully entitled 'A Guide for the Preparation of Aircraft Maintenance Documentation in the International Aerospace Maintenance Language', it was compiled by a special project group set up by the AECMA Documentation Working Group. The Simplified English Project Group is not exclusively European, comprising 15 members drawn from Italy, Germany, France, The Netherlands, the United Kingdom, and no fewer than 13 members from the United States of America. This large American contingent is possibly a reflection of a tradition dating from the 1960s of research and development into technical writing, and the value that is placed there on the importance of producing effective technical handbooks and manuals. The project group, as claimed in the introduction, have 'researched the procedural texts in maintenance manuals' at the instigation of the aircraft industry, and proclaim their prescription for the problem identified by the industry:

In the Aerospace Industry, the airlines identified the need for clear communication of complex maintenance data. Thus, in the late 1970's, the Association of European Airlines asked airframe manufacturers to investigate readability criteria for maintenance documentation within the civilian aircraft industry. . . . This Simplified English is unique in the Aerospace Industry for a number of reasons: it chooses one word for a particular idea or action, defines the meaning that word shall have, and gives a set of rules to simplify the writing style. (AECMA 1989)

Clearly, the authors felt it necessary to encourage the production of maintenance manuals which reduce the risk of any misunderstanding on the part of maintenance crews across the world, most of whom they expected would be native-speakers of languages other than English. They explain it this way:

The user of a manual whose first language is not English may have difficulty with the complexity of English language. To help overcome such difficulties, we have made a set of rules to make the written message easier to understand. . . .

Words were chosen for their simplicity and relationship with other languages. For example, 'occur' is deemed by the committee to be more international than 'happen' and was chosen for that reason. Their strategy can be seen to follow a simple tripartite approach:

1. describe the grammar on which the writing should be based,
2. provide a limited vocabulary comprising a lexicon of about 1400 words, including inflections and variant forms/lemmas (e.g., 'decrease/decreased/decreasing'),
3. lay down individual writing rules for each of the words in the lexicon.

Throughout the document the approach is prescriptive, brooking no arguments and making plain the terms on which the writer would have to take the prescription, as is explained in the introduction:

In Simplified English, a word may have a restricted use. 'To fall' for example, is used to indicate the idea of gravity, and not the idea of a

> decrease in quantity. So the expression 'the pressure falls' is no longer available to the writer who follows the Simplified English rules. He must write 'the pressure decreases'.

This confident and assertive tone is reflective of the whole document, although not all technical authors or engineer-writers are convinced. Without an explanation of the criteria used to decide upon the membership of the list of 'simple' words, they will probably remain dubious about the efficacy of a Simplified English approach. Also, the lack of support for claims about the 'simplicity' of the grammar and criteria for deciding on the 'international' status of words may persuade doubters to remain unconvinced. So many technical writers contributed to the Simplified English project that it seems a pity that more of a rationale and justification for their guidelines has not been provided. Some of the claims are contentious and appear to be based on hunches, like, for example, that their set of rules will make the message easier to understand, or that it is possible to compile an international set of words. This all begs for more detailed explanations and an empirical basis, but the quotations included so far in this section more or less reflect the extent of the justification provided in the rather brief introduction to the Simplified English guidelines. This lack is a noticeable shortcoming considering the intended readers of the document are technical authors, who are usually (highly) articulate in English and skilled composers of technical texts of various kinds.

In the last decade or more, debate has been rumbling on in technical writing circles about the usefulness of Simplified English and other 'controlled languages' (CLs), notably in the 'Communicator', the journal for the Institute of Scientific and Technical Communicators (ISTC). Judging by a recent article, these 'languages' are still not widely used and, it is suggested, unless senior management are fully committed to them and prepared to resource the implementation of a 'CL solution' properly, they are doomed to failure (O'Brien 2005). In the case of Simplified English, the pro-camp continues to argue the case for a strictly limited vocabulary and grammar to reduce ambiguity and serve the needs of those whose first language is not English. An anti-brigade remains sceptical, critical of the constraints it places on writers and the way it hinders innovative writing. The fact that proponents still find it necessary to argue their case seems to indicate that applying Simplified English rules is not so simple, and it is clear that its adoption is less widespread than had been intended. Amongst the teams of technical authors and engineers I have encountered, there has been a reluctance to follow the Simplified English guidelines for writing manuals. There may be a need to reconsider its remit, since, as Kirkman observes about such attempts to restrict language (1992: 136):

> Of course, it is necessary to use a complete range of English to express complex or abstract ideas. Controlled English **can** cope with commercial and technical information like installation instructions, maintenance and repair instructions, operating procedures, and various types of descriptive writing. Controlled English can **not** cope with theoretical discussions, arguments about data, or with very abstract analyses.

Thus, Simplified English may be justified for some text-types, but if it is underused, it may be that it does not serve the purposes of immediate users, the writers, well enough. The whole emphasis of the Simplified English document is on the product and the reader; scant attention is paid to writers, the writing process, or explaining how the documents should be written within the constraints specified. Also, it has to be said, with English now being a global language, that there may be fewer readers in need of simplified English these days. They may be a dying breed.

The case of Simplified English is one worth watching. It is possibly another example of procedures being ignored or disregarded. There is clearly a wish within the engineering fraternity to control and constrain, as evidenced by the content and intent of the procedures. However, in some cases, it seems that producing the procedures somehow spells their death knell. It is as though their significance lies elsewhere, in engineers' professional development, perhaps: the dynamism that goes into the writing of them ends with their publication, with the procedures themselves falling into disuse.

2.2.3 Information Technology Security Evaluation Criteria (ITSEC)

The ITSEC document provides guidance on the writing of technical specifications, and is another example of engineers' belief that problems should be pre-empted or prevented by producing procedures and guidelines, including those for writing. It is also possibly another example of a writing procedure that is ignored more often than not. It concerns the drawing up of security evaluation criteria for special software packages that handle classified data, and the need to develop 'harmonised criteria' across the engineering sector. Unlike the authors of Simplified English, the authors of the ITSEC document are all

British, being members of a certification body sponsored by the British government. The following extract contains definitions of terms used, specifying how the writers of the ITSEC document have used the terms and how readers ought to interpret the guidance given:

Para 0.12
Within the criteria certain verbs are also used in a special way. *Shall* is used to express criteria which must be satisfied; *may* is used to express criteria which are not mandatory; and *will* is used to express actions to take place in the future. Similarly, the verbs *state, describe* and *explain* are used within criteria to require the provision of evidence of increasing levels of rigour. *State* means that relevant facts must be provided; *describe* means that the facts must be provided and their relevant characteristics enumerated; *explain* means that the facts must be provided, their relevant characteristics enumerated and justifications given. (ITSEC 1992)

The advice given about modal auxiliary verbs conforms with a general understanding in engineering about how these verbs should be used in specification documents (see Chapter 5 for more on this). However, advice in the extract about the verbs *state, describe,* and *explain* shows the kinds of problems engineers face when trying to comply with standards. *State, describe,* and *explain* denote different categories of technical description that engineers are expected to be able to produce in response to this standard. There seems to be an assumption that they know the differences between these and can write accordingly. The problem is, though, that most people would find this difficult. Producing such descriptions would be predicated on knowing the difference between a relevant fact and a relevant characteristic, being able to distinguish between relevant and irrelevant characteristics (and facts), and knowing how to write justifications. To some, 'write a justification' could mean 'write a few brief words of explanation', but to others it could mean 'present a formal argument', a page or so long. The guidance is not clear, and the fact is that many engineers do not know how to follow it.

2.2.4   Customer-imposed writing constraints

It is not unusual for engineers to write to a template provided by the customer, especially in the case of complex project proposals involving two or three companies (or more). A recent proposal that was submitted to the American Department of Defense is a case in point. Preparation of the proposal document demanded international collaboration, involving three American companies and one British in a complex interaction very similar to that detailed by Van Nostrand in his discussion of the defense R&D community in America, where the relationships between the interactants are described as being 'essentially rhetorical and insistently pragmatic' (1997: 142). The proposal had to be written to meet the writing criteria and strict writing constraints imposed by the customer, and all the texts making up the different sections of the document were stored at a protected Internet website, hosted by the Pentagon, which could be accessed by all those working on the bid.

This tendency of customers to specify how a document should be structured and written is becoming more common, reflecting an unfortunate trend in documentation production generally, which extends beyond engineering into other domains, including academe. One of the less desirable effects of this may be found in large proposal documents. Chapters 7 and 8, which show how proposals are structured, refer to lengthy sections, commonly located towards the end. Such sections are usually longer because the customer has specified that they be so. With large projects, bidders have to respond to a Request for Information (RFI) or Request for a Quotation (RFQ) which states in detail how many sections should be presented, how long they should be, the kind of information they should contain, and even the size and style of the font to be used. From the customer's perspective, it is considered important to specify the information they need to reach a decision and how it should be presented. Also, there is a perception that these formats have been devised to both lighten the reading load for the customer, and make it easier to check that criteria have been met. How much easier the task should be is the thinking, if the listed criteria can be 'ticked off'. The whole idea is to speed up the decision-making process and assist with choosing the winning proposal.

However, whether the customer is getting the best or most creative solution by insisting on these standardised formats is doubtful. It is somewhat contradictory to expect creativity to be given full justice within highly prescriptive and restrictive formats, which these templates usually impose. By their very nature, they are anti-creative. There is as yet no evidence that standardising documents in this way has the desired effect. If anything, it could be argued that they are, in fact, making the task more difficult and are encouraging less innovative submissions: customers are more likely to receive lexically dense texts, in which information has been compacted to such a degree they make for arduous reading. Engineers spend much time condensing what they have been told to include, even though they consider some of it unnecessary, in order to meet word limits. This could make it more difficult to differentiate between the different bids, so that readers rely more on aspects of price, programme, or other less design-related aspects.

Research journal entry: Risk-averse writing and outdated writing practices

I've just interviewed Philip, who's made me think about traditional working and writing practices. Philip used to work as a ship's engineer in the Royal Navy, but he is now a field engineer with this company. His peers regard him as a good, clear writer and he tells me that his superiors in the Navy thought likewise. He remembers being conscious of his position relative to others placed higher or lower in the Navy hierarchy, and of being required to write reports, since it had been decided by those 'higher up' that he was quite good at it. Any report he wrote was a solo effort, and had to be presented to his superior officer, a Lieutenant Commander, who would examine it and, to use Philip's own words, 'cross the "i's" and dot the "t's" as he was wont, and then he would put his name to it and say "yes, I agree", before it would then go off for the final signature from the captain himself.' If his superior officer decided the report needed improving in any way, he would annotate it, send it back to be changed, and the process would be repeated. When the Lieutenant Commander was happy with the revised report, it was then sent to his superior officer, the Captain, who would read it, and, possibly finding something amiss with it, would write comments about it, and send it back to the Lieutenant Commander, who would send it back yet again to the ship's engineer. As a result, the report could travel up and down the chain like a yoyo until the highest ranking officer was completely happy with it.

Certain engineering companies are like this. They exhibit features of a more accountable, less trusting, sort of blame work culture that existed in this company in the 1980s, but now the atmosphere here is much more supportive, innovative and fast-moving. It is interesting that traditional working practices have become fossilised in other companies, where, for example, engineers still communicate with each other very formally, writing to each other, even when they're in the same room! Jason and Mike mentioned the problem they're having with Bath, who are running the project they're working on. I've heard quite a bit lately about how differently they run things in Bath, with their old-fashioned working practices and rigid hierarchy. It has close links with the Navy and employs a lot of ex-Navy personnel, they say, so this may be the reason. The Navy is renowned, nay, notorious, for this.

2.3   Collaborative writing practices

This section describes a particular approach to writing as an exemplar of typical writing practice, and to illustrate engineers' writing behaviour and attitudes towards writing. It explains the special role played by the straw man in collaborative writing activity, and compares it with another metaphor, kite flying, to show the strong associations certain engineering texts have with risk-taking and anxiety. This account is based on the assumption that a 'flatter' organisation of the company, and the work culture it encourages, enables engineers to work more closely together and share the burden of the writing load with other members of the team.

The working environment in some modern engineering workplaces has been deliberately arranged to facilitate closer working and writing practices. Open plan offices, lack of physical (and hierarchical) barriers, and team-working encourage members of a team to consult more frequently with each other than in organisations which are more hier-archically or atomistically structured. What this actually means is that any major writing project, for example, the need to prepare a proposal or produce a technical handbook, is a cooperative effort involving collab-orative writing practices, so that the burden of the writing load is shared amongst the team.

2.3.1   What do engineers write with other people?

The email survey described in Chapter 1 included an attempt to gain a rough idea about the kinds of texts that involved engineers in collab-orative writing practices. It was a survey carried out in the early stages of a research project, when it was becoming clear that engineers found certain texts-types more demanding than others, namely, specifications and requirements, proposals, and executive summaries. Certain writing tasks are solo efforts, like the writing of internal administration-focused memos and emails. However, it was thought, texts requiring engineers to consult with each other about their composition must be intrinsically more complex and consequential. They must have, the thinking was, a noticeable impact on engineers' work for them to share the responsibility for writing them. Therefore, one of the questions asked respondents to list any documents they wrote with colleagues. All document categories that were mentioned are as follows, with the percentages in brackets representing proportions of respondents who had mentioned each particular document type:

1. Proposals (33 per cent)
2. Software design/requirements/specifications (25 per cent)
3. Reports (23 per cent)
4. Manuals and handbooks (5 per cent)
5. Presentations (5 per cent).

Fewer engineers mentioned the following texts and documents: comments in code, plans, and manufacturing instructions, appraisal reports, patents, publicity brochures, and training material. In the event, proposals are ranked highest, having been mentioned by 33 per cent of the respondents, demanding the most collaborative writing effort in the company concerned. One engineer explained the proposal writing process thus:

> We tend to write proposals as a group but not doing the same bits together (we do individual parts which are brought together as a whole).

His is a simple explanation of proposal writing, a complicated process that is both time-consuming and costly to the companies generally. An examination of proposals and an analysis of writing done by engineers is provided in later chapters. This chapter, however, examines a particular example of collaborative writing, a systems manual, with the aim of revealing the nature of the task for those involved, the type of communication processes and practices they took part in, and the language they used to produce the document.

Research journal entry: Work practices from the ark

Jason also mentioned how different their problem-solving methods were at Bath. If he has a query to make of a systems or hardware engineer here, he simply leans over and chats to him, or he strolls across the room, pulls up a spare chair and discusses the problem straight away. In Bath, however, they still work in functional groups, e.g. all the softies [software engineers] together, all the hardware people together, all the systems folk together, all the commercial people together, etc. In other words, it's how this company used to be in the early 1990s, when Hardware was in A Building, Softies in B building and so on. Well, if the stories going around here are anything to go by, in Bath you would compose a memo and email it to the relevant engineer, even though he may be sitting within yards of your desk. That just about sums up the difference between the two companies, which is a direct reflection of completely different management styles. If I were an engineer, I know which I'd prefer.

2.4   Risk-taking and the textual straw man

The straw man as catalyst in trust-based writing practices

I would like to borrow metaphors from engineers' work talk (and writing) to illustrate the cooperative attitudes inherent in some of their working and writing practices. In discussions with me, engineers have expressed surprise that I should find straw men interesting. It did not take very long to discover that a straw man in an engineering environment is a very different beast from the academic straw man, that it is not made in order to be knocked down, but rather that it is a malleable creation which serves as a constructive and often catalytic function in the writing process.

Ask academics about straw men, and they will usually speak dismissively of them, believing them to be artificial constructs based on spurious evidence, as one of my colleagues said when I asked him about them:

Ah, that's when someone sets up a case and then demolishes it when there's really nothing to demolish.

Straw men usually have negative connotations in academia, where the term is used somewhat pejoratively. They usually occur in the genre 'academic article', and are derided by journal editors who see them as being specially contrived constructions, created by authors in order to be knocked down, so that they can justify putting forward their own research and theories. In other words, academics may deliberately construct a straw man in order to destroy him, at the same time trying to persuade the reader, in this case, the journal editor, of the need to publish their work. In doing so, they are trying to demonstrate the existence of a niche in their research area, albeit a contrived one, mimicking a well-established writing strategy in academic articles, as revealed by Swales' research into genres (1990: 140–143). There appears to be a fundamental difference between the academic's and the engineer's straw man. I see the former as being an illusion, with destructive connotations, whereas the latter has substance and is essentially constructive.

One engineer, rather taken with my interest in the engineers' straw man, likened the academic's straw man to a Trojan horse, because the intention is for the straw man to slip through without being detected. If it is successful in getting past the editors, it means the academic has convinced them that his writing is a valid piece of research. The engineer explained that the engineers' straw man, on the other hand, is intentionally sacrificial, because they do not expect him to survive intact. Instead, they fully expect parts of him to survive, and for him to serve as a catalyst for the production of something much better. A key difference is that straw men are an accepted and useful part of engineers' work practices.

Engineers' straw men and flying kites

British engineers refer to straw men when talking about text, and find the term unremarkable, explaining it is commonly understood amongst engineers in different companies across the United Kingdom. It is not unusual to overhear references to straw men in snatches of conversation, for example: '. . . can you come up with a straw man for us then, Patrick . . .', and '. . . any sign of that straw man yet, Dave?' They are also used in written communication, as these extracts from two memoranda show:

> Please complete the enclosed spreadsheet by the end of August and I will collate inputs and produce a straw man which we can discuss at our next session, date to be advised in due course.

> The recommendation is that a straw man ILS ARM&T plan should be prepared based on this assumption and including the requirements of Annex B of the ITT.

> [ILS – Integrated Logistics Support; ARM&T – Availability, Reliability, Maintainability, and Testability, ITT – Invitation to Tender]

Engineers recognise a straw man as being a distinctive document that performs a particular role in the construction of text. Basically, they see it as a text that is intended for reshaping. It is composed in order to be changed, with the concomitant aim of involving members of the team in its metamorphosis. Flying kites, on the other hand, signals risk-taking. I have overheard engineers' easy reference to kite flying when they talk about constructing straw men, although kite flying is mentioned infrequently. When asked about the significance of 'kite flying' to text construction, engineers explain that they use the term less often, and furthermore, that they rarely flew kites. A possible reason for this is that flying a kite in text represents risk-taking and unilateral action, which most engineers do their best to avoid. A technical author's view of kite flying serves to further illustrate that engineers view texts very much as they view the products they design, that is, texts can be engineered. His words portray texts as physical objects, having proportion, shape, and substance. Text is malleable, parts of it are moveable, and it is dynamic, its dynamism accounted for, in part, by the risk-taking it accommodates:

> A crashing kite need not bring down the straw man. To the engineer, a straw man represents a whole document, even if incomplete, whereas flying a kite is risk-taking in a particular section of it. The kite may be a single point or idea, and could be brought down by the other members of the team, and crash; however, in doing so it would not destroy the straw man, which is, to quote a worn but nevertheless apt cliché, very much greater than the sum of its parts.

Put simply, a kite is a part of a text which is flagged, usually by the one who creates the straw man, as a part of the text that needs checking or special attention paid to it, because it contains a new or unusual idea. Someone who is putting up a straw man to the rest of the team may, in the course of discussing it, say: 'Chapter 3 is more or less as the customer requested . . . but I'm flying a kite in Section 3.7', meaning 'Section 3.7 wasn't planned for. I've put it in because I think it's needed, but I could be wrong, so have a look at it and see what you think.' Another technical author tried to explain how kites function in writing:

> Say I was to write a straw man for something. I might say this bit about the technical solution: I'm really flying a kite there – it's part of the straw man but it's the risky bit. Could be inaccurate and needs really good looking at by others, specialists. So suppose we had to

produce a rapid proposal, a straw man, perhaps we could indicate in the straw man where we were flying kites...where we weren't sure of our ground.

It is generally thought to be much easier to compose something for engineers to read and to comment on than it is to expect them to produce ideas in writing on a blank page. As one engineer put it: 'you use a straw man to draw a response from people, or provoke them into responding'. Another, a technical author, explained that sometimes a lead engineer may take a minute or two to quickly write down his thoughts about a document for the author to use as the basis for a straw man. Authors have told of cases where a few bullet points on the back of an envelope, capturing flashes of inspiration, have been ample for them to take back to their computers and, to use their words, 'bash out' several thousand words to produce a straw man proposal, say. During this 'bashing out', a figurative reference to the authors working frenziedly at their keyboards, they use a combination of sections 'lifted' from other documents, boilerplates (generic text segments on commonly recurring topics, designed to be used in a variety of document types), and the authors' own verve with words. In this way, a rough but ready document may be presented to the project team to provoke the required effect. One author explained why it did not matter if the document was far from being up to scratch:

> If you put something in front of them, it could easily be wrong, but it's so easy for them to react to it. You want to force them into doing something. That's the problem you have: nobody will do anything until you knock something into shape for them and give them some idea of what it could look like. They sit there struggling over how to start, and this is what kicks them off, and they say 'oh YES!'

Document production is associated, more often than not, with dynamism, busy teams, concentrated creativity, and writing to tight deadlines. Straw men can play a useful role in hastening the writing process. However, they also have a part to play in more sedate writing processes, as the following section describes.

2.4.1   Collaborative writing case study

The previous section portrays straw men as provocative documents, provoking reactions from engineers that will vary according to the quality of the 'straw' used. Whatever shape the straw man is in to begin with, however, does not really matter, since it will inevitably result in sections of text being produced to flesh it out until a reasonable draft version emerges. This draft document then undergoes an iterative editing process to produce the final version.

Using a straw man to draft a system manual

Most usually, a technical author produces a straw man to be used at the earliest stages of document production, although the task may fall to an engineer with more specialist knowledge. This section describes a particular instance when engineers used a straw man to draft a system alignment manual (or handbook), where the management of the emerging draft was passed to a new member of the team. It shows how the straw man functions in collaborative group writing activity, involving the team in a cooperative effort to produce the document. It includes a description of the different stages of the writing process; the writing behaviour of the engineers involved; the construction of the text; and the metalingual/discoursal comment that resulted, that is, the written annotations and spoken comment that engineers produced as a result of seeing the straw man. Ten distinct stages can be identified in the development and production of this manual, as shown in Figure 2.1.

1. Customer requirement for a document

Typically, a company receives a request from a customer for a User document. In this particular case, the company receives a memorandum from a customer, asking for an extra manual to be written and providing a description of the purpose of the manual and why it is needed. The memorandum provides requirements, expressed in general terms, for example:

> There is a need therefore for what loosely could be called the 'Inexperienced Maintainer's Guide to Alignment, . . .

and more specifically expressed demands, for example:

> I would suggest that a simple Block Diagram of the alignment chain would assist in explaining what each stage of alignment is achieving, . . . where there is margin for either mechanical or electrical misalignment, and how it is corrected or catered for, and the effects on the equipment of corrections applied.

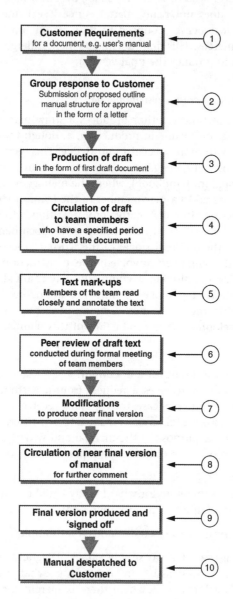

Figure 2.1 Using a straw man to construct a manual

2. Planning a written response to the customer

The team of seven engineers meets to decide how to respond to the customer's request. The team leader has brought to the meeting a straw man to kick-start a discussion on the structure of the response. The straw man has been constructed by a technical author attached to the team. Since a support engineer, who is new to the company, will be given responsibility for overseeing draft developments of the document, the technical author withdraws at this stage to work on another project. He will reappear towards the end, when the document needs to be 'finished off' before sending to the customer.

They read through the straw man, argue about sections of it, brain-storm ideas, and discuss various points, eventually deciding what the alignment guide should consist of in terms of chapters and sections and finally agreeing its overall outline structure. Having decided on the shape the manual should take, the engineers agree the team leader should draft a letter to the customer describing the proposed structure and projected cost.

3. Production of a draft

This involves fleshing out the straw man. Some time later, after the customer has sent approval, a support engineer called Andy is nomin-ated to act as co-ordinator and to produce a first draft. This sees the straw man being transformed to conform with the description put to the customer in the proposal. As co-ordinator, Andy has the ultimate responsibility for seeing to the satisfactory completion of the docu-ment. He is a new member of the team, having recently joined the company, and describes himself as being 'the new boy on the block'. The other engineers see the writing task as an opportunity to induct him into the team: composing the first draft is to assist in a team-building process. The draft still has straw man–like qualities, in that there is still 'straw' needing to be discarded, and this means Andy consults with other team members to replace it, or flesh it out, with text of substance. Andy works almost exclusively on this for about six weeks, consulting with colleagues from time to time. He says the task of producing it has brought him into a closer working relationship with the other members of the group. Another reason that emerges for choosing Andy is his limited knowledge of alignment systems: it is thought he will have a better idea than the other engineers of the needs of the target audience, which the customer had referred to as 'an inexperienced maintainer' (at least, this is the reason given by the other team members for giving him the job of writing it).

Andy writes the first version aiming it, as he describes in his own words:

> ...at the young PO [Petty Officer] who's just finished his course and hasn't really got a clue, and he's dumped onto this Type 23 frigate and one of the operators has come back and says well I think there's something wrong with the alignment. What shall I do about it PO? And he stands there waiting for an answer.

4. Circulation of draft to team members

Andy sends a paper copy of the draft version to each member of the team. They have a week to read the text closely and write comments on the text itself and in the margins. Although they are well used to working (and writing) on computer, this is their preferred method of reacting to the text, that is, annotating 'hard' rather than electronic copy.

5. Marking up text: engineers initially react by annotating

The other members of the group meticulously read and mark up the text, writing comments about a variety of aspects, for example typing errors, missing words, disagreement about lexical choice, and vetoing information that may be misinterpreted or which may have the potential to cause problems for the reader, or, for that matter, the company. Their annotations, and Andy's response to them, are listed later in this section.

6. Peer review of draft text

The review of the draft is conducted at a formally convened meeting of all members of the team, who have with them their own copies with all their annotations. During the two-and-a-half-hour meeting, the text comes under intense scrutiny and comments are totally focused on textual aspects, including text structure, the information content, and language used. The tone of the meeting is formal and matter-of-fact.

As mentioned earlier, the proposed structure for the handbook was presented to Andy when he was nominated co-ordinator. He was aware that the other team members would be involved in the revision process, and that he would have to take part in follow-up meetings with, as he put it, the potential for disaster, for he has been steeling himself for their possible negative reactions. Having presented the first draft of the alignment document to the rest of the team, he now has to meet with them to hear their reactions. This is how he describes the meeting:

> ... but it was explained to me in the first place ... 'The technical part of it ... alright, if it's wrong then we'll shuffle it round 'til it's right.' At no time was I made to feel 'what a load of rubbish that was, go away again, 5 out of 10, see me later' type of thing. It was more of 'OK, we'll collate all the information, we'll have a look at it. Anything we don't believe is correct, for whatever reason, we'll discuss, we'll sort out, and we'll come back again.' They were all part of the team.

From his point of view, the others have an investment in the manual as well, and so they are prepared to share their experience and expertise when it needs to be checked over for the accuracy of the professional and technical information it conveys.

7. Suggested modifications implemented (or not)
After the meeting, Andy collects everyone's marked-up copies and withdraws to consider the suggested revisions. He implements most of them.

8. Circulation of near-final version of the manual
Andy arranges for the near-final version of the manual to be circulated. His colleagues take turns to read the copy and make final comments. These are minor, and so Andy speedily prepares the final version with the help of a technical author.

9. Final version signed off
The team leader gives the document a final check through before 'signing it off', in other words giving it the company's official seal of approval. This is formally entered in company records as having been done.

10. Manual despatched
A previously agreed number of bound and professionally printed copies of the manual are despatched to the customer. However, paper-based manuals are becoming rarer in these days of electronic versions of user-guides and so-called on-line 'helps'.

Text annotations – different types of comment:

As mentioned earlier, each member of the writing team marked up the draft Andy had circulated to them. After the peer review, Andy highlights the comments he accepts as valid and needing to be acted upon, and disregards others he considers to be unimportant. From time to time, he responds to the notes his colleagues have made with his own comments (and occasional retorts) in the margins, as shown in the following examples.

There are three main categories of response from the members of the team, relating to: (1) information and technical content, (2) style, and (3) metalingual comment. Examples of the engineers' written responses are listed as follows and, as can be seen, concern mainly the technical information content of the manual.

1. *Technical and information content:*

   a. **Queries**, for example, *Do you mean SAT 1 (G)? - It is not against an object.*, to which Andy responded: *No I don't I'm referring to DOA test.*

   b. **Criticisms**, for example, comments written against some diagrams about sensor benchmark dimensions: *You have described in detail the STU but not the HATS or SATS. There is inconsistency in the description. It is not clear what the HATS and SATS do from this DOC - AND which one should be used.*

   c. **Identifying oversights**, for example, *We'd agreed to ditch this part.*, and *Explain why not the* MLD.

   d. **Suggestions** for additional material, for example, *Statement on:- (a) How to select page (b) Units entered (c) How to use the page? May be useful.*

   e. **Indicating misleading or inaccurate information**, for example, *Wrong.- you have confused verticality/tilt with racking.*

   f. **Discourse structure and text layout**, for example, *Needs a bit of a lead in to this.* and: *←Is there a heading missing here?*

2. *Style:*

   a. **Criticisms**, for example, *[change] +ve -into positive, not +ve - this is a document, not notes.*, and *←poorly phrased. I know what you mean but this isn't it.*

   b. **(Pernickety) Refinements**, for example, *CAPS? Missing line?*, and *Why unfortunate? To discover any misalignment, you must use the Ref scope.*

3. *Metalingual comment:*

   An apology for over-zealousness, for example: *OOPS Sorry missed this. Thought it was T23 - Delete Comments.*

Research journal entry: Peer comment and a bit of psychology

I discuss with Arthur Edwards a huge handbook he has on his desk. He says that it is only a part of the whole. He mentions how, in his experience, engineers will go to town on a text if they're asked to comment on it, i.e. if they think it is an incomplete document. However, if they are presented with the finished text, they will accept it like lambs. If asked to comment, they will be extremely critical, although Arthur accepts that the comments are usually fair. The point he seemed to be making was that his colleagues can switch their critical powers on or off, depending on whether or not the document is presented to them as a draft or fait accompli.

2.4.2 Evaluating the straw man: engineers' opinions

Twelve engineers who were interviewed about writing tasks talked about the usefulness of straw men in document creation, making the following points:

1. **They encourage team working**, bringing people together to meet and work collaboratively to produce a document.
2. **Straw men force decision-making** within time constraints. In the decision-making process, the straw man acts as a catalyst and has dynamic connotations for engineers. As one engineer put it:

   > So we put up a straw man and we discuss an area. It's got some good stuff in it but it's not quite right, it doesn't quite suit. That forces them to make a decision because it can't go in like it is, we've got to make a decision what do we really want?

3. **Straw men help to identify key points** that need to be made, even though the ones suggested by the straw man are the wrong ones. It doesn't matter. In proposal writing, for example, someone may compose a straw man with the key themes (selling points) as he sees them; he may highlight the fact that the product is fully compliant as being the most important theme. However, on being examined by the team, the straw man may draw out other opinions about what the main theme should be. Being totally compliant may be considered less important than, say, the product already successfully being used by other customers, and therefore having a proven track

record. Alternatively, it may be argued that the after-sales support and maintenance should be highlighted, and so on.

4. **The straw man is the documentary equivalent to a very dry run**. Engineers like to try out equipment to test its design before it is manufactured. This sort of practice with a prototype is an integral taken-for-granted aspect of an engineer's work. Straw men are seen as being similar to dry runs, and the shaping of the straw man like modelling an argument.

5. **The straw man saves time and effort** – As one team leader put it:

> If you have a group of people who turn up to a meeting to discuss a particular subject and a lot of people come along, some of them highly paid, if there's nothing on the table then you have to go through a brainstorming session and then you'd have to get something out before you can get a reasonable discussion going. . . . Now if we hadn't put up the straw man [talking about the alignment guide] and we went to the meeting, we would've spent all day trying to discuss it and really you end up with this sort of ping-pong effect for lots of the time.

6. **Straw men have positive connotations** – The prevailing attitude towards straw men is fundamentally positive. They are regarded as playing an integral part in the decision-making process, encouraging problem-solving discussion, and helping engineers arrive at a solution. Underlying this process is the belief that criticism must be constructive. In the use of straw men, there is no room for vindictiveness or personal attack. Here is how one engineer explained this particular point:

> An important point when you put a straw man up is that people mustn't be offended if other people criticise it, tear it apart, offer constructive criticism. You've got to take it in the light that other people have different opinions. It saves lots of time, because if you hadn't written it down [as a straw man], they wouldn't have thought about it, or some points would have been missed altogether.

Also, engineers like straw men because they are a sort of stress-reliever, helping to reduce the burden of writing long and complex texts. As one engineer explains:

> It is easier to write a straw man, because it doesn't matter if you get it wrong. In fact, you expect to get it wrong, so you can rush it out

any old how. It is safe to do so because you know it will be sorted out later.

2.5 Summary

Treading the fine line between caution and risk is the balance engineers have to achieve. For as long as people's lives depend on the effectiveness of their designs, it will always be so. The painstaking attention they pay to methodologies and work procedures is understandable, although it seems these hamper more creative aspects of their work at times. It is indeed a problem for them to work with the conflicting demands of bureaucratic-like controls and the need to be innovative. Scientific methods are their mainstay, though, providing them with tried and tested methodologies to follow, and it is these that form the basis of their approach to most tasks, including the writing of documents.

In the next chapter it will be seen that certain documents are on a par with the products they represent, and this is why it is important for engineers to extend the same treatment to these documents and exercise the same controls over them, or at least try to. The documents are substitute products, or in the case of the alignment manual just discussed, an actual product. Therefore, documents receive the same sort of treatment as any other product, with engineers bringing to them the same care and controls. However, documents, together with the texts and language they comprise, have proved slippery for engineers. The language engineers use (including computer codes) has proved difficult to pin down, although it seems they are tireless in their attempts to harness and control it. Their use of English seems problematic, possibly because few engineers have specialist knowledge of its structure. Later chapters continue to examine engineers' special use of English for the various complex tasks they have to perform.

If the previous paragraphs have presented an overly gloomy view, it is worth recalling the straw man and the part it plays in the writing process. Any straw man heralds creative activity, and engineers view them as such. The straw man is a strikingly apt metaphor for any design task and is typical of an engineering approach to text (and the product). Most design work demands cooperative team effort and a team 'spirit' which essentially demands that engineers strive together to produce a product to satisfy a customer. It follows, then, that there is little room in modern engineering for those who are not 'team players'. In the process, the product in its textual form can be reshaped and rebuilt, incorporating any of the kites that manage to keep flying and on the right course. In this way, engineers view texts very much as they view the products they design, that is, as objects that can be engineered. This chapter has shown that engineers working on text together practise an essentially cooperative way of working that is constructive and instructive. Their approach serves as a useful model for those within and without the engineering sector.

3

The Engineering Product

3.1 Introduction: setting the scene

This chapter is about how working engineers view product development. It charts the whole process diagrammatically, in an attempt to set the scene for later chapters that discuss the texts that engineers produce as they communicate at work. Diagrams play an important part in this chapter, because they better portray the complex ideas that surround any product. Engineers like diagrams, often preferring them to words. They will draw on any piece of paper that is close to hand, no matter how large or small, to communicate an idea. The diagrams in this chapter draw upon their ideas to present this story of the product.

'Product' in its general sense refers to anything that results from engineers' work, although in this book it is used more particularly to refer to the entities, or things, that they design for the customer. There is a popular misconception linked to a traditional view that engineering products are concrete things (as distinct from abstract things) with hard shapes and substance, usually referred to as hardware by engineers. This is an incomplete picture, however, as in design engineering, 'product' may be used to refer to any entity that has physical substance or, in the case of software, textual substance. 'Solution' is another word used to refer to the product. Chapter 1 mentions that 'solution' is a customer-centric word used in situations where the customer is present, and in texts that the customer will read.

Engineers are perpetually preoccupied with the product. Their work is so intertwined with it that any fortunes or misfortunes that occur in its development have an impact (sometimes immediate and dramatic) on their working lives. They are knowledgeable about different theories about design processes, and sometimes develop their own. When describing a particular theory about processes, one engineer explained that the design process could be seen to be a gradual reduction of uncertainty, with decisions made at key points that focus engineers' creative efforts down ever narrowing paths. At each of these stages, competing solutions are analysed and the best alternative chosen. The design eventually becomes more detailed and specific until the point is reached when there can be no turning back. If the engineers have followed the right path, the customer is provided with a solution well suited to his needs. However, if engineers follow the wrong path, the customer could find himself with a product that meets his specified requirements but does not meet his needs (Brookes 2005), a situation to be avoided, if at all possible.

Engineers are fascinated by processes generally, and usually have a clear view of how their work contributes to the overall development or use of a product. Engineering text books reflect this interest. Ask engineers to explain product development and they effortlessly provide descriptions closely matching those in the books they have studied.

A wide choice exists of diagrammatic representations of the kind of product life-cycle typified by Jones (1989: xii). Apart from being circular in some way, a striking feature of such diagrams is their commonality with process diagrams in textbooks for other subjects. For those working in education, more specifically, educational technology and syllabus design, very similar process drawings are available. This would seem to indicate that a basic cyclic diagram well represents in a general way the life-cycle of any 'product', be it a physical artefact, computer software, or an educational course. Inevitably in an engineering environment, an examination of product development involves studying diagrammatic representations and, in this section, a somewhat different view of the product is put forward to the ones usually found in engineering text-books. After all, idealised process diagrams do not always tally with actual goings-on.

3.2 The product life-cycle

3.2.1 Knowledge accrual in engineering design

Each engineer is usually attached to a particular product or group of products, has a good idea about why it is required by the customer, and knows about the various stages in the design and making of that product. Those not involved in early design work can refer to the documentation, which plays a key role in recording all design, business, and contractual decisions taken from the outset. A range of documentation is produced during the life of a product, and this section shows various depictions of the product from the earliest stages of its conception as an abstract idea through to the very end, when it is no longer used and is discarded by the customer. This cycle may last for as long as 20 or 30 years in the case of some products, for example the wing of an aircraft or a gun-firing system for a frigate.

Engineering and business books provide numerous examples of diagrams representing production and commercial processes, although none could be found that relate engineering design processes to the documents they generate, with the exception of Thomas *et al.*, who show five stages in the development of a product, together with associated document types: user requirements documents, software specification documents and design documents. Showing document types in a diagrammatic context is useful indeed, although their linear arrangement of the stages, with a clear beginning and end, does not really match reality (Thomas *et al.* 1994: 18). For this sort of development work, circular diagrams, like the one developed by Jones (1989: xii), are more appropriate, because they show the process as being cyclic and comprising stages. Jones typically identifies 'seven distinctly different phases of a product life-cycle', although these are wholly product-oriented and labelled to express the cycle from a product perspective, so that 'pre-concept' means 'pre-concept of the product', 'full-scale development' means 'full-scale development of the product', 'disposal' means 'disposal of the product', and so on. However, such diagrams present only part of the picture. If a diagram is to be of any use at all to this study, it needs to show four key elements: the human, textual, temporal, and product dimensions of the story. Since a search yielded little information about these four elements in diagrammatic format, it was decided that some should be drawn to better suit our purposes.

For the cyclically inclined, Figures 3.1 and 3.2 were developed to portray not only the part played by the product as a dynamic process, but the roles of the engineers and customer as well. The product, as a concept or physical entity, is a key concern at every stage. It is a constant element throughout, whereas human participants change, as do the types of documents. Figure 3.1 shows that design engineers, who receive more attention in this book than others, are involved only in the design of the product, and the diagram shows in some detail the stages of product development which are their concern: these range from early thoughts about the product to the making of a formal proposal about the product to the customer. The former may be captured in scribbles (literally) on the back of envelopes, day- and log book entries, technical

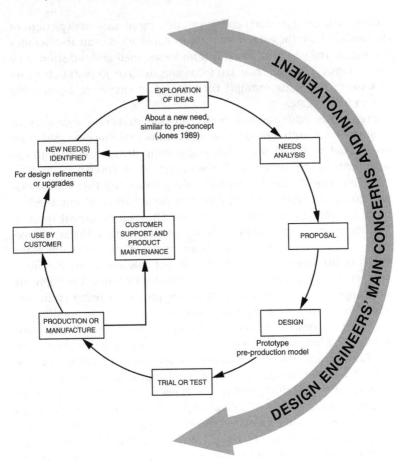

Figure 3.1   A cyclic representation of the design process

notes, and Requests for Information (RFIs); the latter may be in the form of lengthy text, submitted formally to the customer in bound volumes. Figure 3.1, then, is an attempt to present a cyclic depiction of design engineers' work.

However, a more accurate representation of the product life-cycle is a spiral. A spiral conveys the dynamic and continual nature of the design process. It spirals outwards in an ever expanding growth, to reflect the accrual of knowledge and expertise that builds up as the product is designed, tested, and used. Thus, a product life-spiral can be seen to evolve out of Figure 3.1 to show the huge growth of knowledge that can result. It also reflects the major impact of small decisions taken in the very first stages on the course of events that follow. As an engineer explained it: 'small decisions in the early days wipe out a whole range of possibilities, narrowing down the options to only one outcome. That's why the early design work is considered so important.'

All the knowledge that accrues is stored in some way or, to use the engineers' term, 'captured', most usually in orthographic form and engineering drawings, but even quick sketches and scribbled notes in an engineer's daybook would be considered part of the body of knowledge that is built up. More recently, this knowledge has been captured in electronic (as distinct from paper) format in word processed files and electronically held 'notes' and 'comments'. Whichever the format, however, this body of knowledge is held in text or diagrammatic form, and stored or 'saved' on computer as something precious, sometimes immeasurably precious, to the company and to the engineers. Figure 3.2 conveys a sense of knowledge accrual, a growing body of knowledge expressed as design data or design information. As the design develops, it becomes both more specialised and more detailed and intricate. The narrowing of the focus enables more and more detail and therefore more

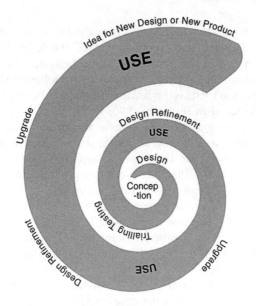

Figure 3.2 A product life-spiral

information to be generated. Figure 3.2, then, shows the expansionist nature of the information being produced. However, since the temporal dimension of product development has received little attention so far, we now attempt to include this important dimension in the discussion. This needs to be adequately accounted for in any diagram that aims to present an accurate portrayal of engineers' work.

3.2.2   Temporal considerations in product design

A fundamental temporal consideration: products may have a long shelf-life

A good illustration of the significance of temporal matters in product development is the gyroscope. It is a product that was originally used to assist in navigation, although it has been designed and redesigned and now has a variety of uses. An engineer, who was talking about the success of recent gyroscope-based products, explained that they had evolved out of an original invention made over a hundred years ago by Elmer Sperry. To quote his words:

> The momentum of the business stems from that original design. We are where we are today because of Elmer Sperry [who designed the original gyroscope].

The gyroscope he was referring to was originally conceived in the late 1800s, with a formal design proposed in 1902. The design for the mechanical, rotating gyroscope formed the basis of navigational equipment used by the British Royal Navy. A ring-laser gyro followed this mechanical gyroscope, and then (as it is usually referred to) the silicon gyro (SG), which was designed in the early 1990s using silicon-chip technology. The SG looks set to form the basis of new navigational aids for missiles, and more recently as a motion sensor for the automotive industry and two-wheeled vehicles like the Segway Human Transporter. It can be seen, then, that the so-called 'life-cycle' of a product can extend across several decades, or more. In the case of gyroscopes, it has extended across the whole of the twentieth century and into the twenty-first. Long timescales are common in engineering, particularly in the defence industry, and this is a significant factor when considering engineers' writing (and working) practices and the potential longevity of design documentation.

Figures 3.3 and 3.4 attempt to show these long periods of time that may be involved in product development. Figure 3.3 is, in fact, a pared-down version of Figure 3.4, showing the three main categories of

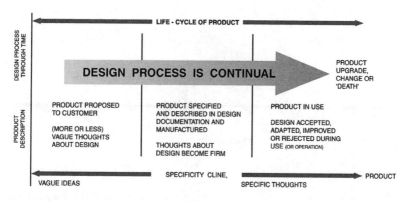

Figure 3.3   Simplified product life-cycle: three main phases

documentation that specifically relate to the product. It can be seen that these three categories coincide with three macro-phases in the product life-cycle.

These phases are described in more detail later, but at this point it is worth noting that they coincide with distinct phases involving more, or less, interaction with the customer. Where there is more interaction with the customer, in the first and third phases, for example, the documentation is written for customers as the main audience, and with engineers having a clear focus on the demands made by them. The documentation produced in the middle phase, however, when they work on detailing the design and manufacturing the product, is intended mainly for internal audiences, comprising colleagues within the company, review teams, and other collaborative companies. The customer is rarely concerned about documentation produced during this central phase, which, in turn, has a fundamental effect on writing practices and text structure. Chapter 5 explores in more detail the writing and other communication activities relating to this phase.

Figure 3.3 also shows a parallel process of design refinement, which is indicated in the diagram as a specificity cline. This represents the fact that engineers may begin with (sometimes exceedingly) vague ideas which are shaped through text to produce a specific design, and then, ultimately, the product itself.

Figure 3.4 shows the key stages in the life-cycle of a product, together with the main participants and document categories. It attempts to convey a sense of the temporal dimension and to provide a

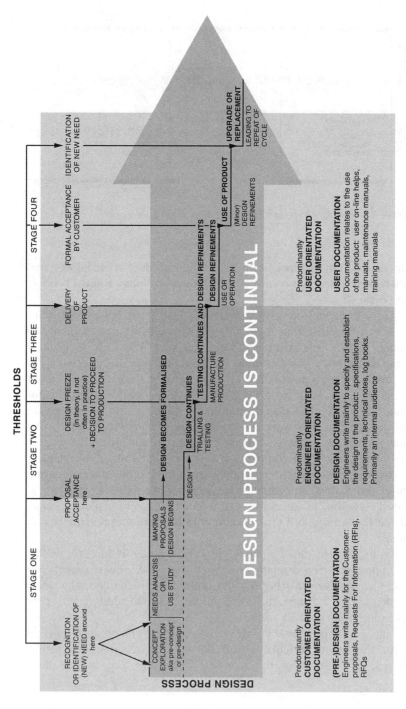

Figure 3.4 A text-oriented view of the product life-cycle

historical perspective in which to locate main activities, processes, and document types. The spiral in Figure 3.2 is incorporated in it, although it has of course been unravelled to better show the chronological ordering of:

- Key stages and events
- Main work processes
- The focus (or 'orientation') of communication tasks at each stage, that is, whether the texts are produced for an external or internal audience.

Unlike the compartmentalised models of the kind discussed earlier, it depicts the design process as being a continual one. Its portrayal of the temporal dimension shows the whole period of time the product is in use, and even potentially beyond that. It is possible for a product to 'exist' for many years, if the technology is not superseded, making it obsolete through further design developments. It shows the stages of the 'life' of a product from an engineer's perspective, rather than from a finance, or marketing and sales perspective, and, overall, provides a realistic context for considering the documents (and other spoken and written texts) that engineers produce. It is, if you like, a meld of my applied linguist's perception and engineers' views of the life-cycle of a product.

Figure 3.3 and particularly Figure 3.4 are intended to serve as reference points for the consideration of communicative events in the engineering workplace. Most importantly, they provide a visual context for the consideration of engineers' texts, which are a product of communication tasks. The significance of having such a textual-oriented view of the product is beginning to be appreciated by the engineers themselves, as there are indications that some recognise that texts have intrinsic value. In particular, product-related design texts have a special long-lasting value, containing information immeasurably valuable to those working on engineering design. One document manager refers to this as a 'soup of information' and he is currently working on storing all the design work produced by engineers in digital format, including all drawings and written texts. Although a mammoth task, this data capture is essential for any modern information-driven organisation. It is, in effect, the capture of the products themselves in an electronic textual store or 'warehouse', where they remain for years or decades, only to be erased whenever the product they represent falls into disuse, or is beyond repair. Although he does not use the terms 'text' or 'language', it is this type of textual representation of the product that Kidd is referring to when he observes:

> the configuration, or information, life-cycle, is longer than the product life-cycle. It starts with the first definitive set of information, and ends with the retirement of the last information package. (Kidd 2001: 38)

3.3   A textual perspective of the product life-cycle

In this description, thresholds are like boundaries, the point of entry to the next stage, and have no temporal significance. In real life we aspire to achieve certain goals, and 'threshold' may be the name we give to the metaphorical 'place' we strive to reach. Once we get there, the time it takes to 'cross' a threshold may be negligible, in marked contrast to stages, which can last for weeks, months, or even years. Generally, decisions are reached at a threshold (or actions performed), which clearly push the engineer onto a new stage and into work with a distinctively different purpose. For example, the customer's identification of a need, which is the first threshold, leads engineers into a prolonged period of 'blue-sky' thinking, when they talk and write together to generate new, innovative ideas. This then leads to an intense writing period for engineers, during which proposal documents are produced. In this way, work on the product moves forward, stage by stage, ensuring its progression and evolution into new stages of development or use.

As Figure 3.4 shows, six thresholds have been identified and labelled in terms of activity focus, and a brief description of the stages is given, mentioning whether the documents at each stage are primarily for internal or external consumption. The usefulness of making a distinction between customer-oriented or engineer-oriented documents may not be strikingly obvious at this point, although it becomes more significant when considering the 'anxiety'-index of documents. This factor is particularly relevant to the writing of proposals, specifications, and requirements, discussed in later chapters. The main work activity of engineers at each stage is also shown in Figure 3.4, since it is assumed that the communicative intent and rhetorical structure of the documentation will reflect this activity. During Stages 2 and 3, for example, engineers work on design documentation and formalising the design in text. These thresholds, and the stages they bound, are described in the next section.

3.3.1 Thresholds and stages in product development

The following is a list of the thresholds and a description of the stages they trigger:

Stage One: Needs Analysis and Proposal

Threshold One – Customer's recognition, identification, or description of a need for a product.

As Figure 3.4 shows, this threshold leads to a detailed exploration by engineers of the customer's need, sometimes referred to as a needs analysis or use study. The distinguishing feature of this stage is that it is predominantly customer-oriented. Discussions take place with the customer that result in documents being produced for him to see. This stage particularly involves the production and interpretation of high-level specifications. It is the customer who usually specifies first, while the engineers respond by drawing up a proposal, which the customer ultimately accepts or rejects. This stage is not always instigated by the customer, though, as there are occasions when engineers take the initiative to point out to him that he may need a new or updated product. Texts that are typical of this stage are (a) high-level specifications and requirements, (b) proposals, and (c) RFIs (Responses to Requests for Information), which more logically should have the acronym RRFI, but do not, for some unknown reason.

Stage Two: Design and Testing

Threshold Two – Customer's acceptance of a proposal submitted by engineers.

Usually this should mean one of two things: in the case of a competitive tendering exercise, the proposal is the winning submission, and in the case of a non-competitive bid, the customer gives the go-ahead to the company for its engineers to move on to the next stage of product development. In both situations, usually, the contract is signed, which means that any work the engineers do is now funded under contract, with expenses now covered by the customer rather than the company. The engineers can now move on to design a proto-type or pre-production model, and test it. Trialling and testing are key activities during this stage, playing an integral part in establishing the ultimate design of the product. A distinguishing feature of documentation at this stage is that it is produced mainly for the engineer's own use, and so can be described as being predominantly engineer-oriented. Documentation at this stage is concerned with detailing the design and testing of the product.

The previous paragraph reflects the theory put about by textbooks, the ideal situation. It is also the line trotted out by the engineers, until they are challenged about its depiction of real workplace practice. Then they will readily admit that in practice they rarely have the luxury of testing a proto-type. These days it is not uncommon for them to work with the customer in developing the product in the field while it is being used. Engineers also admit that both the engineers and the customer maintain a pretence, which starts when Threshold one is crossed, that the product has already been developed or is well on the way to being so, paying lip-service to the new procurement procedure brought in by the MoD in the early 1980s under Margaret Thatcher. It is not unusual for a similar fiction to be maintained with civil contracts. Apparently, this is a common phenomenon throughout the industry. More recently, however, procurement procedures have been introduced at the MoD to better reflect current design and production processes.

Texts at this stage centre on specifying refinements in design and, in writing them, engineers aim to produce exact and clear descriptions and definitions. Since they need to write about a myriad of product features, the texts can be long and exceedingly detailed. Contractual specifications, which typify engineers' work here, are 'ticked-off' to show the customer's requirement has been satisfied. 'Ticking-off' in this context refers to the placing of ticks in boxes, usually found in special sections at the end of documents, when specified tasks have been completed. It is done for the customer's satisfaction, although engineers also have a sense of satisfaction and achievement when this is done. 'Ticking off' is a formal ritual that marks the end of this stage.

Stage Three: Production

Threshold Three – Decision that the design and testing have been completed for the purposes of production.

By crossing this threshold, engineers are able to see to the manufacture of the product or, in the case of software, its installation. However, at times, this particular crossing proves a difficult one for them.

Design is little different from writing: with certain products, engineers will return again and again to certain aspects, fiddling to get them absolutely right. Refining a design is like honing text, in that each time it is revisited an improvement here or there can be made. There comes a time, however, when this refinement must stop. In the design process, this particular point is called a 'design freeze'. The idea behind this is that it should bring to a halt any more tweaking of the design. If there is such a phenomenon as a design freeze, inevitably, it must be asked: Is there ever a 'design defrost'? Interestingly, there apparently is (and often). Engineering theory will have it, though, that all known needs should have been identified by this stage, and any problems should have been solved. Any change in instructions after the freeze usually involves a formal review of the contract, bringing with it complications in terms of design revisions, redrafting documentation, and arguments over costs. However, projects can be dogged by unforeseen problems, which may be caused by the customer changing his mind, or by engineers failing to foresee design implications in the original specification. It is this stage which has the potential to be the engineers' worst nightmare: the product which was a figment of their imagination has to take on a form and substance, and, more importantly, has to work. When it does not, huge losses can be incurred, and a product may end up costing a company much more than was originally anticipated. This is the nightmare scenario that engineers will do their utmost to avoid.

Documents produced during the production stage are almost exclusively generated within the company for internal consumption, and can be described as being predominantly engineer-oriented. Since actual practice does not match theory, it follows that further refinements to design documentation may be made. In the case of one product, a gun system mentioned below, further specifications were produced with the benefit of hindsight seven years after the so-called 'freeze', expressed in two weighty documents.

Stage Four: Operation or Use of the Product

Threshold Four – Delivery of the product to the customer (who may or may not be the user).

Engineers focus particularly on customers and users at this stage, particularly users, although they may not be the customer with some products. For example, a customer, a government department, say, may commission the design and building of an aircraft, but he will not fly it. The users in this case would be the pilots and other flight personnel. The main concern during this stage is how well the product functions in the field, and, especially, how well the user works with it. For the purposes of this discussion, 'customer' is taken to mean 'user' as well.

It is believed that crossing this threshold leads the engineer to move into a distinct new stage in which the customer plays a key role. In

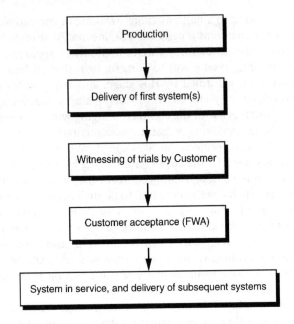

Figure 3.5   Ideal post-delivery scenario

an ideal world, armed with the original specification, the engineer spends a year (or two or three) developing the product, undisturbed by the customer who patiently waits on the side-lines. And, in fact, the ideal scenario is very occasionally realised in actual practice, so that the sequence shown in Figure 3.5 may occur. However, Figure 3.5 portrays a traditional view that, in truth, rarely happens. Engineers talk about 'best practice' these days being 'to manage the customer's expectations throughout the design stage so that he really understands what he's going to get, and what it will and won't do, before he gets it'.

With certain products, the customer is involved in the trialling of a product as soon as it is delivered, and provides feedback to the engineer as part of a continuous post-production design process. The documentation for a particular product, an automated gun system, for example, reveals that it was actually in service for six or seven years before it was finally (albeit reluctantly) 'accepted' by the customer. This product proved to be a huge problem to both company and customer in terms of cost and use. Although such long-delayed acceptance is unusual, it is common for such products to be in use at various places around the world for about a year before the customer formally accepts them. This enables the products to be monitored over a period of time to see how they function at different latitudes and climatic conditions.

Distinctive activities at this stage are in-service customer support, product maintenance, and servicing. As a natural consequence of these, much of the documentation is user-oriented. Texts being used or developed at this stage are manuals, handbooks, and other textual assistance for the user, for example on-line helps.

Stage Four continued: Use of the Product

Threshold Five – Formal acceptance of the product by the customer.

In the case of certain products used by the military, this is called the Fleet Weapons Acceptance (FWA).

This threshold represents a part of the contract that applies to the product after it has been used for an agreed period. Put simply, this is the stage at which the customer officially states that the product has been produced according to the contract and operates satisfactorily. The idea behind this formal acceptance is that it relieves the company (and engineers) of the burden of responsibility for any unforeseen problems that may arise subsequently, and saves the company from having to bear extra costs that may occur as a result. Clearly, a line has to be drawn at a certain point in the development of a product so far as product support and maintenance is concerned. It is only reasonable for a point in time to be reached when the customer can no longer expect to draw upon the engineers for help, maintenance, and repairs, at least not without paying extra for this.

Once this threshold is achieved, the customer continues to use the product, and through extended use and changing circumstances, may have ideas about how it could work even better. He may even identify new problems and new needs, or the engineers may do so, which then leads them to the next (optional) threshold.

Threshold Six (optional) – Identification of new need.

This leads to a re-run of the whole process, with the engineers and the customer working, usually in consultation with each other, on developing a refinement, up-grade, or totally new product.

3.4 Summary

So far, a picture has been developing of engineers at work, who they are, what they do, and matters they find important to their work. The first three chapters of this book have been concerned with providing a backdrop, or context, for considering particular texts and communication tasks that arise in later chapters. Apart from the engineers themselves, this context comprises essentially four main elements, the four 'P's, if you like: engineering processes, procedures, practices, and products. These influence engineers and everything they say or write to an extent that many outside engineering fail to appreciate.

This chapter, then, completes the setting of the scene. It shows the product as being central to engineers' interest and activities at work. The path of product development is rarely smooth, but the checkpoints along the way, the thresholds described in this chapter, provide engineers with some kind of assurance and stability in what can be a complicated and difficult process. The discussion conveys a sense of the importance engineers (and the customer) attach to exerting some kind of control over the whole product development process, especially since the product may be a highly complex system.

This is why engineers attempt to follow procedures and processes in product development. Their continual search to improve ways of working is simply because of their wish to best harness their creativity within set time-frames and cost constraints to meet the targets that have been set. From the outset they need to establish a clear direction to follow in order to become more focused on the product. In the course of narrowing their focus, they produce increasingly detailed descriptions of the product, generating more ideas and knowledge that need to be taken into account. Understanding this side of the story of product development goes some way towards helping us to understand the complex demands made of engineers when they communicate. These demands, and the texts that result, are discussed in the following chapters.

4
Engineering Texts

4.1 Introduction: engineers' view of texts

This is a transition chapter. Previous chapters discuss the engineers as a distinctive group, their attitudes, work practices, and the customer's role. Later chapters examine the texts that particularly concern them, describing the structures and language used to express these texts. Product development is a constant element throughout, being the primary focus of engineers' work activities and the communication tasks they need to perform. This chapter uses information gained from interview data and results of the survey described in Chapter 1 to discuss texts through engineers' eyes, beginning by describing what engineers write, and drawing upon information gathered from engineers about their views on text categorisation and document types. It will be seen that engineer-derived categories vary somewhat from received opinion in applied linguistic circles and the usual portrayal in English for Specific Purposes textbooks that focus on writing technical reports and instructions.

The chapter also describes engineers' attitudes towards the texts they produce and views about the language they use. It develops the idea raised in previous chapters about the product being the central focus, describing engineers' views about how the product should be portrayed in their documents. It will be seen that they do not always agree on certain points, especially aspects of persuasion, and that they have a tendency to produce (and favour) formal, complex language.

The change of focus in this chapter paves the way for the study of texts that engineers consider to be special in later chapters. Of all the different types of document they have to write at work, it will be shown that only two are of real concern (and interest) to them: proposals and specifications. In fact, these are complex 'matrix' documents or colonies (Hoey 2001: 73, 93), comprising different texts-types and genres, with spoken and written spin-offs requiring different communication skills, both written (engineers' main interest is in written communication) and spoken. This chapter provides a bridge to these special documents, by explaining engineers' attitudes towards the texts, the kind of information they need to communicate through them, and the specialised nature of the language they use.

4.1.1   Formal versus informal

First, let us consider the meaning of the terms 'formal' and 'informal', as an engineering slant on these is somewhat different from the one taken within applied linguistics. The terms tend to be used rather loosely whichever the domain, but, generally speaking, linguists use them to reflect the context associated with a text. Desultory chat among acquaintances in a pub, for example, would be considered informal language, whereas a parliamentary debate or senate hearing would produce language considered as formal. Engineers, on the other hand, tend to use the terms 'formal' and 'informal' to refer (it must be said, inconsistently) to different notions. Chapter 5 explains one notion that concerns the distinction between what engineers refer to as 'natural' language and a more controlled 'formal' language of specification and machine coding. In the case of documents, however, much depends on whether or not they are open to scrutiny by external agencies, that is, anyone 'outside' the company. Engineers tend to regard documents that have external audiences (and this tends to be the customer in most cases) as formal, and documents intended for internal audiences as informal. Later sections show, however, that engineers tend to use formal language when communicating with their colleagues as well.

Although linguists would recognise texts as being formal by virtue of the strict procedures followed in the writing of them and in the formal style of the discourse, ultimately, to engineers, any 'formal' label for a text denotes income generation. Formal documents usually require the approval of the customer before they are deemed complete and having achieved their purpose. They demonstrate that the company has jumped through particular hoops in the development of a product or service (see Chapter 3 on the life-cycle of a product). When the approval is gained, the customer makes a payment to the company, which is usually a staged payment. It follows, then, that engineers tend to regard as informal any document intended for an internal audience, even though formal procedures are followed in their composition and the language used is every bit as formal as that used for the customer. As with any flexibly applied term, there are exceptions: engineers are influenced by hierarchical connotations, so that a document issued by senior management for an internal readership, or a document intended for senior colleagues to read, may also be referred to as 'formal'.

4.1.2   Ephemeral versus 'lasting' texts

A distinctive feature of any discussion involving engineers is the graphic illustrations that usually accompany the talk. Drawings and diagrams of all kinds, be they created on computer or on the back of an envelope, are a common feature of an engineer's professional communication. Whether discussing something in passing, at a formal meeting, or at a presentation, the need to accompany the talk with a quick sketch is a natural part of the discussion. Others have described this feature of scientific texts (Davies and Greene 1984, Myers 1990, Halliday and Martin 1993), showing the integral role diagrams play in conveying meaning. Most of the diagrams in this book have been devised either to communicate my ideas to the engineers or, as in the case of Figure 4.1, for them to relay their ideas to me.

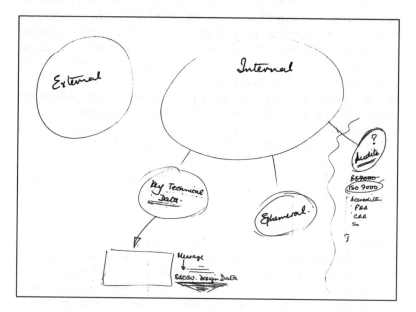

Figure 4.1   An engineer's depiction of text categories

Figure 4.1 was drawn very quickly to illustrate a senior engineer's view of the texts engineers produce at work. His drawing shows two broad categories: those generated for external audiences and those for internal audiences. The largest category is the internal one, which comprises three sub-categories: key technical data, a major category, and of most significance to engineers; ephemeral texts, which he described as 'admin-related texts of no lasting value'; and documents produced for audit purposes. Whilst adding embellishments to the drawing, he commented that the 'key technical data' category is of greatest interest and concern to him as an engineer, notwithstanding his management role.

I was struck by his reference to ephemeral texts, and inferred from his gloss of them that engineers considered them unimportant. To me, this was potentially useful information for approaching an investigation of engineers' texts, my reasoning being that the textual field could be conveniently narrowed by eliminating those text-types considered to be of less importance. Later, questions were included in the email survey to help identify these texts, with unexpected results, as will be explained.

One of the main aims of the questionnaire was to identify those writing tasks that engineers were most concerned about. There was anecdotal evidence that engineers found particular tasks exasperating and questioned the value of what they were required to write in certain cases, for example SOFT reports ('SOFT' being the acronym for 'Successes, Opportunities, Failures, Threats'), and other writing required for internal administrative purposes. Since engineers used uncomplimentary epithets on occasion to refer to admin-related texts, it seemed reasonable to infer that ephemeral texts, by virtue of having brief lives, had no lasting value, and were therefore a category of text valued less by engineers. It also seemed reasonable to assume that engineers may consider such texts unimportant, much as professionals in other fields regard as tiresome some of the internal administrative writing they have to do (Davies and Forey 1995), and especially academics at universities, who are finding increasing demands for admin-related writing becoming onerous and value-detracting.

In the event, the engineers' responses revealed the kind of breakdown in understanding that can occur when an Arts-oriented person, working in the Higher Education sector (notorious these days for bureaucratically inspired writing tasks), tries to communicate with engineers and scientists, whose writing is almost entirely product- or customer-focused. The engineers' (occasionally indignant) responses to the survey question motivated me to take issue with the engineer who first coined the 'ephemeral' category of texts. The ensuing discussion revealed that, although ephemeral texts have no long-lasting value, engineers do not necessarily consider them unimportant. The word 'unimportant' is too value-loaded and emotive for their liking. It means quite simply that, unlike design documentation, which endures, ephemeral documents are relevant only for the time they serve any useful purpose. Often the purpose is served within short periods of time, possibly days, or even minutes, after which they are no longer required and no longer kept.

Research journal entry: Another angle on ephemeral documents

Saxon is yet another engineer who has an angle on modern versus outworn writing practices. From his perspective, there are 30+ engineers miles away in Birmingham who are very good at generating lots of paper and reports, but aren't nearly as good at design work and are achieving very little, whereas in this company (again, according to Saxon) just he and two other engineers are cracking all the code, as it were, and helping to get the folk in Birmingham out of a mess. Unfortunately, the project is being managed by Birmingham. The auditors who visited from there recently were not impressed by the lack of administrative documentation produced by the engineers here, and slapped them on the wrists. It is clear from listening to him that his story is a good example of 'ephemeral' documentation being considered trivial or not as important as design-related work. I know he'd baulk at 'trivial' and 'unimportant', but that's how it seems to me. Since the company here has cut down on the amount of paper engineers have to generate for administrative purposes, it must have been rather a shock for them to be dragged back to such old-fashioned practices.

4.2 The documents engineers write

4.2.1 Engineers don't write anything they don't value

It appears that most engineers are in the fortunate position of being required to write only those documents that they value. A noticeable feature of writing produced by engineers today is that little relates to administration or bureaucratic matters, the vast proportion of written output being concerned with engineering design. A majority

(66.66 per cent) do not write anything they consider to be unimportant. This contrasts with findings of other surveys (Davies and Forey 1995, Davies, Forey and Hyatt 1999: 298–301). Eleven engineers simply wrote 'All' as an answer, when asked which types of writing were important to them. Others felt the need to explain themselves:

> This doesn't answer your question, but I consider them all important. If they weren't important, they wouldn't need doing, and I'd find something more important to do.

> In some way, they are opportunities to put your mark on something permanent, and a way of getting known about the company. This can of course, backfire in a big way!

> They all have their importance, some not at the time that one wrote them, e.g. engineering log book entries.

> ?at the risk of being flippant – all of them – they are all part of the job

This situation may be due to fundamental changes brought about in certain engineering organisations and work practices that have enabled a greater proportion of engineers' time at work to be spent on value-added activity. To improve their competitiveness, companies are seeking to provide organisational structures and physical environments to better facilitate engineers' almost total concentration on customers and products. Such changes had been made around the time the questionnaire was administered: the company concerned had undergone a reorganisation and rationalisation of work practices, so that certain administrative type documents, SOFT reports and the like, were no longer required, as the following answer reveals:

> Nothing [is unimportant]. There was a time in the past when we had to write what I call Processed Junk, e.g. soft reports. But they've done away with that now.

These changes to work processes have had noticeable effects on the writing engineers have to produce, reducing the range of text-types to be written and causing engineers to be more motivated about writing them. It could be argued that their belief in the relevance and importance of their writing is proof of the efficacy of streamlining and improvements to work practices that have taken place in the company concerned. This has resulted in engineers only writing what is important and useful to their work, and, it logically follows, the writing being more focused (on the product).

4.2.2 The product is the focus

It has already been established that, when engineers are communicating at work, the product is their main concern. It follows, then, that the most highly regarded writing tasks are those which are essentially technical and design-centred. In other words, the texts they value most describe the product and are essentially descriptive, although they are distinctively different from each other in terms of intended readership and purpose: requirements, specifications and software design, reports, proposals, engineering log book entries, executive summaries, and presentations. The most valued from this list are:

1. Requirements/specifications/software design
2. Technical reports
3. Proposals and bid documents
4. Engineers' log books.

Engineers spend most of their time talking and writing about the product to make clear aspects of its function and structure, manufacture and assembly. One feeds into the other. Engineers who are designing different parts of the product need to communicate design descriptions to other engineers, who see to constructing and testing various aspects. Thus, any technical description usually defines the product, defines terms, and explains what it is and what it does. Text for this may be explanatory, informative, or even educative, containing information on how it works, naming main parts (shape, size, similes relating parts to everyday objects), and what they do.

It can be seen, then, that technical descriptions vary according to their function, and this function in turn is determined by the type of product being described and the audience for whom it is being described. Taking the customer to be the intended audience, let us consider, as illustration, the fundamental difference between commercial off-the-shelf (COTS) products and bespoke products that are designed from scratch. A silicon rate sensor, as an example of the former, is an ingenious component with general applicability, useful in a range of commercial, military, and space products. Potential customers, however, may not know much about the product or that it even exists, let alone how it functions, how it is structured, or how useful it could be. These customers may need to be taught about such aspects, through specially composed descriptions.

These descriptions aim to persuade potential customers to consider buying the product, and in the event of them doing so, they need descriptions to advise them how to use it. Whether trying to persuade a prospective customer of the merits of a design, discussing with the customer's representatives the performance of a prototype, or writing a product specification, engineers need to produce a variety of technical descriptions about the following aspects of the product.

1. *Functions of the product: describe what the product actually does when it is in use*, mainly through written (and occasionally, spoken) description. They need to be able to explain how the product is useful, which means they need to describe the different functions it performs. In simple terms, they need to explain what it does, and this is quite distinct from explaining how it does it, which is also part of a functional description. 'Function', 'functional', and 'functionality' are words used rather imprecisely by engineers, and are difficult to pin down, not unlike the situation in linguistics.

2. *Appearance, structure and construction*:

   a. *describe the appearance, structure and construction of the product* through photographs and diagrams (galore), and accompanying text. Engineers are design-conscious, in that they appreciate and aspire to produce elegant functional designs. They acknowledge the primacy of functionality and usability, although they are nevertheless adherents of fundamental design principles, which demand simplicity of style, economy of scale, size, and materials, and ease of use. Such descriptions may also include detailed descriptions of the product's construction, the materials it comprises, its component parts, and how it is manufactured.

   b. *provide a system description*, and, depending on the genre or document type, feature the product from a particular perspective. If the product is part of a complex system, it would usually be described in two sections dealing with electronic and mechanical aspects.

   c. *describe aspects of production* and how elements of the product should be manufactured (in production specifications), and provide details of the factory assembly of it.

   d. *describe the design concept (or the basis of a blueprint)* and this would include 'blue-sky' (or down-to-earth) thoughts about design. Some technical descriptions form the basis of the blueprint of a design ('blueprint' taken to mean 'model' or 'plan'). As the response to an enquiry by a potential customer, say, or as a solicited bid, a technical description may be sent to customers who are thinking of commissioning the design and production of a bespoke or new product. Thus, such text may be used at the earliest design, or rather pre-design, stage.

3. *Performance*:

   a. *describe the product's performance* using graphs and diagrams, and provide accompanying explanations of the mathematical calculations and scientific principles to substantiate their expectations of the product's performance. This sort of description is commonly found in proposal documents, reports, or marketing flyers, where engineers need to explain what the product is capable of doing, provide information about how it has performed in tests, or describe how it has performed with other customers who have bought it. So engineers would also:

   b. *describe the results of tests* carried out on the product under certain prescribed conditions, as well as research and development activity on the product.

4. *User-oriented descriptions: describe the product from the user's standpoint.* Such descriptions take account of the user's perspective, for example a user's manual would explain how the product should be used (in contrast, 'selling' documents, like proposals, describe the ease with which a product can be used), how many people are needed for it to function properly, how safe it is for the user, how easy it is to maintain, and how to install it. Also, the description may state what spare parts are available, and how to carry out checking and maintenance. Engineers refer to such texts as 'Support and maintenance' documentation.

5. *Compliance*:

   a. *describe how compliant the product or 'solution' is*, that is, the extent to which the product gives the customer what he asked for (which may not be the same as what he really wants; see Chapters 7–10 for more insights into this tricky notion), and how the design fulfils the customer's requirements. Such descriptions give an idea about degree of compliance, and are persuasive in intent in that they aim to convince the customer that the product would be best for his purposes, even if it does not quite fit the bill (or, to use engineers' words, is not '100 per cent compliant'). This aspect of description is especially relevant when trying to persuade the customer to buy the product in oral presentations, for example, or in written engineering proposals. From the customer's perspective, the degree of compliance may be the most important consideration when deciding whether or not to choose the product. Cost considerations may be important as well, of course.

b. *describe the main benefits* (or selling points) of the product, usually in oral presentations and proposal documents, discussed in later chapters. Benefits are linked to notions of compliance, and are separate points of information about how useful the product is, or could be (if it is being proposed).

Research journal entry: Product descriptions may have a long shelf life

Any text that explains, usually to a prospective customer, what the product does, engineers refer to rather loosely as a product description. Steve Montague [marketing man] calls them this. Product descriptions are often needed by engineers concerned with marketing or 'selling' a product, e.g. an early enquiry by a foreign embassy. Sometimes these requests are made at very short notice and, in Steve's case right now, it is a bit of a surprise, because it's an enquiry about a product that may be coming to the end of its shelf-life. The team tells me it is 'nearing obsolescence'.

So now I see Alan [technical author] hassling Brian Pearson for some information on the SEAL. Brian's dug out something, but Alan thinks it isn't good enough. He's looking for some technical information that particularly conveys the usefulness of the equipment. All of a sudden Brian and the others seem to be engrossed in what's lying on their desks, avoiding Alan's gaze as he marches up and down having a bit of a rant. He's now diving into the backs of cupboards and rummaging in filing cabinets. He clearly finds it ironic (and maddening) that 'there are twenty-four pieces of kit out there on the ships, but nowhere can I find a decent description to put in this RFI!' He's saying all this to bowed heads.

Alan has to have it drafted for Steve by the end of the day. He's finding heaps about the thermal imager, but then they're not enquiring about the thermal imager . . .

[Author's note: this is an early journal entry and, in this particular company, all documents are now stored electronically in a massive computer archive to prevent just the sort of situation Alan found himself in.]

4.2.3 The main text types

Essentially technical and design-centred

The survey into writing practices revealed the wide variety of writing that engineers produce at work. A total of 22 different categories of writing task was identified, which for purposes of simplicity have been grouped notionally according to mode (format), message (discourse function), and target-audience (readership). Reports form the largest category with 80 per cent of engineers writing a range of technical reports. Almost as many write design requirements and specifications (71.7 per cent), including those related to software design. The high number of engineers involved in proposal writing (also 46.6 per cent) is a reflection of the nature of engineers' work activity in recent years, which is often concerned as much with business-seeking, as well as dealing with in-service products. The need to write proposals to 'sell' their product (and their expertise) is not unlike the situation research biologists find themselves in, having to constantly write research proposals (Myers 1990: 41).

A surprise finding was that a significant number of engineers (46.6 per cent) write log book entries, a dark-horse genre, since there is little mention of these in engineers' discussions or in the literature on technical writing. Up to the 1960s, undergraduate students used to be required to practise writing log book entries as an integral part of their engineering courses in British universities. This taken-for-granted recording by engineers of their design development and creative ideas seems to be under-valued, or even overlooked, judging by the meagre attention it now receives on engineering courses.

Fewer engineers (11.7 per cent) seem to write manufacturing instructions or manufacturing specifications, possibly reflecting their reduced involvement with the implementation of a design in the production process, since the trend to manufacture in other, lower cost, countries has become more common. The same number mentioned having to write plans, revealing in their answers that they have to write a variety of them. Plans seem to be regarded in much the same way as engineering log entries, in that they are a taken-for-granted aspect of engineers' work. Plans are concerned with plotting or extrapolating work schemes mainly in diagrammatic form with accompanying written explanations, and, since engineers find them unproblematic, receive little attention in this book. Technical notes are similarly treated.

A small number of engineers (10 per cent) mentioned writing technical notes, which are primarily concerned with capturing a design idea. The small number may be accounted for by the fact that technical notes are not formally linked to any particular product. Since companies usually only calculate the cost of producing documents for particular products, and allocate money for producing these documents, resourcing is not available for technical notes, which may be very brief or an extended technical description of a few thousand words. According to guidelines issued by one organisation, technical notes give 'factual statements of events, calculations and work done' and are intended for an internal readership.

But this is only part of the story. From time to time, the thought may occur to an engineer that some idea he is working on may prove useful to later design work, and he sets about describing it using words and diagrams. Its usefulness or relevance to products being developed may be obscure, but the engineer has a hunch it might be needed in the future. Even though it is intended for an internal audience, a technical note is a document that is formally recorded by the company (in the company library, if there is one), indexed, and stored. This formal recording of the idea is necessary for establishing intellectual property rights (see 7.3.4) and for fighting any counter-copyright claim. Depending on company regulations they may have a restricted readership within the company, being released on a need-to-know basis, since the product (or idea) concerned may have a sensitive security classification. Even smaller numbers of engineers are involved in writing manuals and handbooks, patents, or procedures.

Format-centred, that is, not content specific

Significant numbers of engineers mentioned writing memos, emails, and letters (75, 38, and 30 per cent respectively), but only 15 per cent mentioned faxes, the use of which has declined rapidly to virtually zero since the survey was conducted in the late 1990s. In most cases, the memos are email messages, and can be viewed as being an email subcategory, since a wide variety of texts can be sent as emails. In her more detailed treatment of emails, Surma refers to them as a hybrid rhetorical form, before examining writers' and readers' slippery perceptions of each other during this 'written conversation' (Surma 2005: 132). The term 'email', in fact, denotes both mode of transmission and format of messages, but not the message type or genre. In this respect, faxes and letters are similar to emails, in that they are modes of transmission, and are often similar to those 'letter-format documents' observed by Freed (1987: 158) to be formal proposals produced by an accounting firm. Letters that are transmitted on paper via traditional means using postal or courier services carry a special significance today, since the bulk of correspondence is composed and sent using computers. These days letters are usually of a contractual and/or financial nature requiring careful preparation, often in response to a query from a customer or supplier.

Engineers are emerging from a transition stage, having moved in the past decade from sending mainly paper-based messages to transmitting nearly all messages electronically. It follows, then, that the formal letter is on the wane, and will be limited mainly to those communication tasks with legal ramifications, that may have legal (or cost) significance in the future.

Writing to facilitate speaking in front of an audience

Very small numbers are concerned with what may be described as facilitative writing. This is writing that engineers produce as preparation for presentations they make from time to time to colleagues about new work processes or scientific matters that will impact on engineers' work. Facilitative writing of this kind helps to implement new procedures and aids professional learning within a company. Engineers are also called upon to make presentations to the customer, which may be lengthy events, lasting three hours or more. Writing for presentations, which concerns 8 per cent of engineers, is a special skill, because it makes different demands from other tasks: there can surely be no other writing task that requires engineers to script-write, that is, plan what they are going to say in front of a live audience. The small number of engineers reporting involvement in this kind of writing activity may be due to the time the survey was conducted. The fact is that most engineers consider presentations to be important 'texts', no matter whether they are intended for colleagues or external audiences. At certain times in their working lives, every engineer has to make them, and when they do, they take the exercise very seriously indeed. Oral presentations tend to have this effect on people, no matter what the context. (Chapter 7 includes a case study of an oral presentation to the customer.)

Intended for external audience, essentially persuasive and/or 'selling'

The distinctive feature of texts in this category is that they are intended for external audiences, usually customers or potential customers.

Fifteen per cent of engineers write executive summaries, overtly persuasive sections of proposal, which they find particularly problematic (examined in Chapters 7 and 9). Only one engineer writes web pages, and two engineers prepare brochures and other publicity material. Both web pages and brochures require particular writing skills which are very different from those required for writing the design-focused texts in the first category.

4.2.4 The problem documents

When asked which documents consumed most of their writing time at work, most engineers mentioned reports, specifications, and requirements. The following are examples of responses to this question:

> Probably the specs, as they are laborious to produce, and require frequent changes. Specification writing and responding to customer & supplier questions and comments.
>
> Technical reports, as they are generally linked to applied research projects. These can take several weeks to complete as they evolve with the research. Requirements also take a long time as they will change several times during the review process.

Although writing reports and technical notes is time-consuming for engineers, they do not find reports or technical notes problematic. Specifications, requirements, and proposal writing, on the other hand, present difficulties to them. An examination of mainly face-to-face interview data (Sales 2002) reveals the kinds of documents that trouble engineers most. These results are based on direct references made by engineers to texts and writing, and have proved influential in the direction taken in this book and the decision to exclude texts that engineers do not find problematic. Report writing is a thoroughly researched field, and well described and documented in the literature (Marder 1960, Pauley 1973, Souther and White 1977, Houp and Pearsall 1980, among others); however, not one of the engineers mentioned any aspect of report writing as being an issue, apart from it being time-consuming. Similarly, technical notes, tests records, and log book or (electronic) journal entries trouble engineers too little for them to bother mentioning them.

Engineers who work almost exclusively on engineering design tend to be concerned about winning more business and see writing proposals as playing a major role in helping them to do this. Their livelihoods may rely on writing winning proposals. Figure 4.2 shows that they are mainly

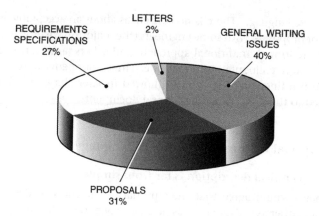

Figure 4.2   Problematic writing

preoccupied with two broad types of document: (a) proposals and (b) specifications and requirements. The majority of the problems raised, a total of 58 per cent, concern proposals (31 per cent), and specifications and requirements (27 per cent). Forty per cent of the problems engineers mentioned relate to general writing issues, concerning writing processes, writing skills, and company writing procedures. The few references to letters were made by engineers who were involved in commercial aspects of preparing bids (Chapters 6–8). A few of the respondents suggested that engineers' inadequate mastery of the English language had caused difficulties in the production of technically related documentation, especially requirements and specifications, which demand precision of expression.

Specifications and requirements are examined in the next chapter, but it is worth mentioning at this point that engineers need to be sometimes suitably vague, or precisely specific, when writing them. Their failure to be one or the other at appropriate times, for example, when writing a technical specification or composing a persuasive section of a proposal, has led to problems for companies. So far as design documentation is concerned, it is generally believed that misunderstandings have arisen through poorly written documents which have proved (exceedingly) costly in time and money (Kincaid 1997: 54).

There is some disagreement about the use of persuasive language, however, with some engineers saying that proposals could be improved through the inclusion of more persuasive language. However, the survey has shown that engineers are not clear about what is acceptable persuasive language. There is no consensus about an acceptable writing style for proposals, with some engineers (the majority) wishing to adhere to a more formal, traditional approach, and others (a minority) advocating a more colloquial, informal one. These comments were to prove useful in influencing the course followed in later chapters, which are devoted to these problematic proposal documents.

4.3   Special features of engineers' language

4.3.1   Technical description is far from simple

Engineers would agree that their primary concern is to describe the product either as accurately or as persuasively as possible, depending on their purpose for writing (or speaking). Everyone thinks they know what the word 'description' means, but it has proved rather more slippery for applied linguists to pin down. Description is a discourse function commonly found in written and spoken language, and is one of the four 'basic types of writing' mentioned by Brooks and Warren (1952, cited in Urquhart and Weir 1998: 83). Davies (1995: 88) refers to terms like 'descriptive' and 'expository' as labels that refer to 'the broad social or communicative goals of the writer'; whereas Martin (1989: 7) refers to description as a genre, closely related to reports; and Grabe and Kaplan (1996: 218) see description as a writing task that is a report sub-category. Basically, work on text categories is still very much 'work in progress' so far as applied linguists are concerned, and they continue to try out different ways of talking about text, text categories, and discourse, so that 'description' tends to be used rather loosely as a cover-all term. If we consider common everyday situations, description fulfils a variety of purposes across a range of genres, as the following three familiar examples show:

1. 'Describe the film' means something like 'summarise the plot', 'explain its message' or 'explain the moral of the story', 'identify its theme', and so on, depending on the context.
2. 'Describe the view' could mean 'provide a word picture of the scenery or landscape, including most striking geographical features, colours, vegetation, and man-made aspects.
3. 'Describe how you spent the weekend', a typical topic for a primary-school writing task that most pupils find loathsome, means 'produce an extended piece of writing in the form of a school essay', a genre rarely found outside the school environment.

Engineers are called upon to produce technical descriptions of various kinds more often than any other text type, and are unequivocal about the kind of writing needed to produce them. They seek to write as clearly and objectively as possible, and look for these same qualities in the writing produced by others. One engineer expressed his view about the need for clear expression thus:

> I often have to rewrite old instructions...which are ambiguous, vague or not understandable. In this form they are ignored by the Operators and we wonder why work is not being done effectively.... Every time I do a re-write, I recognise later on that I could have done a better job of it.

Engineers generally regard any kind of vague language to be anathema, especially when describing the product. They believe it should be avoided at all costs, although, when examples of vague language they themselves have produced are pointed out to them, they will admit it can be useful in the early stages of product design. Engineers are no different from others who need to communicate in a commercial environment, and at times exploit the subtleties of the English language to serve special (persuasive) purposes. Proposal documents reveal numerous examples of engineers using vague language as a hedging device (Channell 1994), simply because the proposed product may still be an underdeveloped (even hazy) notion at this stage. In the following examples, engineers can be seen to be hedging or leaving options open:

- 'The accelerometer currently envisaged', where the verb and adverb imply 'but we may change our minds later'.
- 'Initial mechanical design concepts', a noun phrase with suitably impressive pre-head modification ('initial mechanical design') that possibly serves as a smokescreen for the posing of a tentative idea: essentially the concepts are initial and may change, as the engineers have yet to develop the idea fully.

A longer discussion of engineers' attempts to be persuasive in text is provided in Chapter 6. This section discusses engineers' aims to write clearly, for it is common to hear them saying that their written expression should be 'clear, concise, correct (or accurate)'. This is the kind of advice that is often found in self-help books on technical communication and textbooks for engineers (Fear 1977: 59, Houp and Pearsall

1980: 161, Haslam 1988, Ellis 1997: 161, for example) that is liked and often repeated by engineers themselves.

4.3.2   Complex simplicity

However, producing clear unambiguous language is no easy task. In his, albeit brief, reference to technical language (in his Introduction to Functional Grammar), Halliday mentions the 'often professed ideal of "plain, simple English" '. He remarks on the deceptiveness of this phrase, because 'the concept of "plain and simple" is very far from being plain and simple'.

He explains that any kind of technical language tends to become even more complicated when attempts are made to simplify it (1994: 350), and alludes to the fact that written language, and certainly technical language, has a tendency to be clausally simple. For example, an examination of a collection of engineering specifications, a type of technical description, would typically show that sentences and clauses have ostensibly 'simple' structures, for example:

SUBJECT (S) + VERB (V) + SUBJECT COMPLEMENT (Cs) – an SVC
  structure
SUBJECT (S) + VERB (V) + OBJECT (O) – an SVO structure
SUBJECT (S) + VERB (V) + ADVERBIAL (A) – an SVA structure

The structures are ostensibly simple because the sentences have so-called 'simple structures', which nonetheless contain complexity. They follow SVC- or SVO-type clause structures of the kind shown in the sentences below, all three of which are taken from technical descriptions. They may be deemed 'simple' because, according to Quirk and Greenbaum (1973: 166), they do not have embedded clauses as constituents. However, such sentences may contain structurally complicated noun phrases at S, O, and Cs positions, as the following sentences show:

The combat system designer will incorporate a low risk electro-optical tracking system compatible with displays, weapons and a range of sensors via any ship's highway.

More recent versions of the sensor use silicon, a material with a strength to weight ratio three times that of steel, as its vibrating element.

The heart of any Coriolis gyroscope is the resonator itself with the device performance acutely dependent on the stability of material parameters.

In terms of sentence constituent structure, these may be categorised as 'simple', but the noun phrases comprise several nouns strung together in what Halliday describes as 'a pile-up of nouns' (2004: 159):

- the noun phrases 'the combat system designer', 'more recent versions of the sensor', and 'the heart of any Coriolis gyroscope', functioning as Subjects;
- 'a low risk electro-optical tracking system, compatible with displays, weapons and other sensors via any ship's highway' and 'silicon, a material with a strength to weight ratio three times that of steel', functioning as Objects;
- 'the resonator itself with the device performance acutely dependent on the stability of material parameters', functioning as a Complement.

It is the density of the information compacted into, and conveyed by, these noun phrases that renders these sentences far from simple. Such structures (and ones containing even more complex noun phrases) are a distinctive feature of engineers' writing. This is probably because engineers attempt to be objective and concise within predetermined writing word limits, while at the same time including information about complex notions and mechanisms. Engineers are, after all, fundamentally scientists (or applied scientists) by training. The complexity of constructions, such as 'the device performance acutely dependent on the stability of material parameters', which post-modifies the head noun 'resonator', has been observed by Halliday as being a kind of nominal construction commonly found in scientific writing, and typical of such disciplines as physics or mathematics. He also observes that it is such features that non-specialists find difficult to read (2004: 171, 159). All things considered, it is unremarkable that the task of writing, and reading (Davies and Greene 1984: 42), is a difficult one.

Halliday provides a detailed account of scientific language, explaining why non-specialists find scientific texts (and engineering texts would be included in this category) difficult to read and understand. He puts forward both the specialists' and non-specialists' points of view. On the one hand, he expresses the oft-heard opinion of lay people that scientific writing is unnecessarily complicated and difficult and could be made much simpler and easy to understand, if only non-technical terms and more colloquial English were used (2004: 160). On the other hand, however, he explains that many of the ideas in scientific writing are 'highly complex and often far removed, by many levels of abstraction, from everyday experience', and so it is understandable that scientists, and in our particular study, engineers, are unable to express complex notions using 'everyday' English. As Halliday writes: 'technical terms are not simply fancy equivalents for ordinary words' (ibid.: 161).

4.4 Summary

Anyone coming from an English for Special Purposes (ESP) background would be struck by engineers' radical views about text and the fact that these views are held consistently across the discourse community. To a man (and woman), they all hold similar views about the importance of any text that describes the product. They are chary about using words like 'important' and 'unimportant', considering them too emotive for describing their texts. All the same, they would recognise the primacy of product documentation, in marked contrast to any other administration-related texts, which they not only see as ephemeral, but usually pay scant heed to. It would give a false impression to say they disregard ephemeral documents, because they pay close attention to certain messages at the time they are received. But these are clearly peripheral to their main work and concerns.

They are essentially prospective in their views of texts. This is a thought arising out of Sinclair's comments about prospection (2004). Sinclair's observation in fact concerns 'the prospective features of spoken discourse' rather than written, but the notion of engineers having a forward-looking outlook when writing is an appealing one. When they write, they are prospective, because they know their texts will be acted upon by others in the development of the design. Eventually, out of their texts will emerge the product itself, or they can foresee, for example, the user relying on their texts to operate the equipment. Engineers find these 'prospective' texts the most interesting, and it is invariably the case that these are texts concerned with the product, the customer, and the user.

They are not terribly interested in reports, which is an interesting finding. Report writing has long been regarded as a major course topic, and report-writing skills a key area, in ESP. However, engineers write reports because they are an expected part of the job, and so they write them dutifully but not very enthusiastically. It would seem that this is because engineering reports are essentially retrospective.

The next chapter discusses engineering specifications and requirements, a specialised kind of technical description that is probably the most objective to be found in any engineering documentation. It examines engineers' attempts to represent the product in text , as a precursor to later chapters that discuss the central importance of product-related sections in proposals.

5
Engineering Specifications and Requirements

5.1 Introduction: delivering on a promise

Earlier chapters describe how engineers attempt to capture design work in text. In particular, Chapter 3 shows how the germ of a design idea is often recorded in proposal documents. Within these documents, it is the technical description of the engineers' design intent that provides a reference point for the drafting of requirements specifications. Promises need to be converted into deeds. Engineers need to revisit the proposed solution to ensure the company delivers what was promised and fulfils its contractual obligations. It may happen that, on returning to the proposed solution, engineers actually improve on the original design and find they change it radically. Whatever course they take, the fact remains that the proposal provides the impetus for writing specifications that, in turn, specify the product or products that are part of the solution.

In textual terms, however, there are two fundamental differences between proposal and specification documents: communicative function and cost considerations. First, communicative function is a key consideration; there is no persuasive intent in the writing of specifications and requirements, whereas persuasion underpins every aspect of writing in proposals. The solution put forward in a proposal undergoes a sort of writing metamorphosis that takes place through a series of writing, design, testing, and manufacture phases. Eventually, the solution is transformed into something tangible that can be used by the customer. Using specifications to transform (sometimes vague) ideas into a physical product is the main purpose of this post-proposal stage, with functionality being a key concern. Second, specifications and requirements have financial value and have the potential to generate income for the company. Proposal writing, on the other hand, is an expensive activity for any company, let alone those in engineering, and for the motor, aerospace, and defence industries, the costs are huge. With proposal writing, a company makes financial outlays on document production in the full knowledge that it will have to count its losses if the proposal loses. So, it takes a financial gamble when deciding to bid for a project, and will deem the money to have been well spent, obviously, if it wins. Documents that specify the product, on the other hand, that is, specifications and requirements, are regarded by companies (and engineers) as having monetary worth, because they receive payments from the customer when these documents are produced. They earn income for the company whenever they have been deemed to have been satisfactorily completed in the eyes of the customer. This is a significant consideration indeed for companies, cash-strapped or not. As one engineer put it: 'These are documents that the customer sees, checks against criteria, and pays lumps of money for.' In the case of very large documents, a company may receive payments on completion of each section.

5.2   What are specifications and requirements?

5.2.1   Clarifying terms

Drafting specifications and requirements is a fundamental engineering activity. Whether it is a hardware or software system, the engineer aims to describe its functions and physical features in documents called *specifications* or *requirements specifications*. These are read by the customer and engineer colleagues concerned with designing, testing, and manufacturing the product, providing evidence that the design work, and all the scientific and mathematical testing that this involves, has been done, and done properly. *Specs*, as engineers usually refer to these documents in their talk, are hefty tomes, often comprising several hundred pages, their weightiness a reflection of the engineers' weeks of toil spent specifying in detail every possible feature of the product as clearly and unambiguously as possible.

Specifications, and the more detailed *requirements*, are descriptions of a special technical kind. Put simply, they are an attempt to describe the design for those who will later use the specifications to convert them into the product itself. The *spec* has to be sufficiently precise and detailed for the product to be built and tested from it alone, and this detail is expressed through individual *statements*. Statements may be made up of a single sentence or a paragraph comprising several sentences. The whole purpose of drafting specifications is to write as clearly as possible, although a view circulates among engineers that they themselves complicate the task, through lack of skill in English, as they would put it. The real story is not so simple a matter, however, as this chapter reveals.

Chapter 3 explains how the engineering proposal spawns a myriad of design specifications, if it is successful. In the proposal, the design idea is explained in such a way as to appeal to the customer and to convince him that it is the 'solution' he seeks. If the proposal is accepted, the design engineers need to set to work on writing out the details of the design for others to act on. In the case of a complicated navigation system, for example, these others will be other design engineers who will specify, in a written description, the design of the specific components of the system for yet other engineers to examine and act upon. This action may involve further writing about the manufacture of the hardware components, or designing pieces of software to operate the system.

Without exception, engineers agree that specification writing is a major area of concern to them (Sales 2002). They not only spend much of their working time on them, but have a genuine interest in improving the way they are written. Too many stories have been told of misinterpretations and misunderstandings that have arisen through poorly written (so engineers say) specifications, which have proved costly to their company. From time to time, news items about problems arise concerning a particular design feature, for example that the latest upgraded aircraft are unable to use smart bombs that can be programmed to hit specific targets, very important in an age when the public expects, and demands, that there should be no 'collateral damage', and that civilian homes should not be hit. Engineers follow such news stories keenly. In the case of an aircraft like the Tornado, it is usual for the design specification to be drafted jointly, with the writing of the requirements shared by the company and defence department involved. In their work talk, engineers refer to a 'requirements spec' [*sic*], and invariably refer to the government or country that has commissioned the product as 'the customer'.

Linguists and those outside the engineering community find engineers' language curious (or just plain odd), particularly the way they use modal verbs. That such usage endures to the present day must be due to engineers' views about 'English': they seem to have a low opinion of the adequacy of the English language when it comes to writing engineering specifications and requirements, and have devised their own rules in an attempt to make English serve their needs better. In practice, these rules are not always applied consistently (Kirkman 1992: 125), leading to problems and misunderstandings with the customer and, in the end, it seems these attitudes are ambivalent and somewhat confused, as we shall see (in Section 5.6 on modal verbs).

5.2.2 Specifications are the textual bedrock of engineering

Taking a textual or linguistic view, a specification is a document, which is usually written by the lead engineer of a project. This document is an interpretation of the customer's requirement, and develops the technical solution detailed in the proposal as well. For large projects, there may be several specifications, or documents, each of which concerns a particular aspect, for example physical features and composition, electrical and electronic components, functional features, environmental considerations, dimensional and spatial aspects, and in-service support. For large projects, like the construction of a space shuttle or an aircraft carrier, for example, there may be hundreds of specifications, or documents, comprising millions of requirements.

A requirement specification is, in fact, a variant of the text type 'technical description', examined in Chapter 4, and like any other text will be expressed according to its communicative function. Such texts have a dynamic function by virtue of their utility and the fact that readers are directed to perform particular tasks as a result of reading them. As such they may be referred to as performatives, to borrow (and mis-use) a term from Austin (1975: 148). The target audience will determine the nature and content of these texts: the technical description will vary according to the reader or, to be more specific, the reader–user. So a specification written by a systems designer will start with a general description, which will then be rapidly broken down into individual requirements according to 'how the designer's mind works' and how he conceives the whole breaking down into a multitude of different parts (and apparently disparate components). Specifications specify expected action and behaviours of those concerned with the design and manufacture of a product, since they may have quality, safety, cost, and legal implications for the company. Inevitably, and in view of what could happen (see second research journal entry: *How complete can a specification be?*), engineers need to work on the principle that everything relating to anything put in print (or not, as the case may be) in an engineering document may have legal or cost implications ultimately.

5.2.3 A glut of guidelines, but a dearth of information

A search of any company or university library would yield books on multifarious aspects of engineering, bar specifications, and requirements. The stark fact is that it is unusual to find books on these, and even if they can be found, they tend to deal with procedural matters. The internet, however, abounds with information about specifications, although again the guidance is couched in general or procedural terms, without dealing with the specific problems engineers are faced with when writing them.

In spite of the general acceptance that requirements are a major problem, and that badly written requirements can lose a company millions of dollars, little has been written in engineering literature that directly addresses the problem. There are few exceptions (Hicks 1961, Haslam 1988, Ellis 1997), and the treatment they receive tends to be cursory or of a general nature. It is difficult to find anything other than vague definitions of what they are, or should be. Even less has been written about specifications in the field of applied linguistics with, again, few notable exceptions (Brusaw, Alred and Oliu 1976, Fear 1977, Kirkman 1992). For an industry that is fond of defining terms, useful definitions of specifications and requirements are hard to find.

Established reference books, for example engineering and technology dictionaries and manuals used by engineers, reveal there to be a dearth of information about these terms. The *McGraw-Hill Dictionary of Scientific and Technical Terms*, for example, which boasts in the blurb to be the world's most comprehensive single-volume reference and an indispensable tool for scientists, engineers, students, and the like, has no entry for either 'specification' or 'requirement'. The main library at a British university renowned for its engineering departments had only one book on the writing of specifications, and it related to those in civil engineering (Haslam 1988). There seems to be a commonly held assumption, amongst the engineers and in engineering literature, that requirements and specifications are so commonplace, so fundamental to an engineer's work, that they do not need defining or explaining. However, in view of the difficulties engineers have when writing specifications, it is clear this needs to be addressed.

Engineering companies do try. The writing guidelines issued by one, for example, are typically entitled 'Hierarchy of Engineering Specifications' and list 12 different types of specification. Of the 12 listed, 5 of the categories relate to design. The rationale behind these guidelines, according to the company document, is as follows:

In order to properly devolve the requirements for, and specify the Company's products, a hierarchy of specifications has been developed. This hierarchy ensures that the requirements are captured, devolved and recorded in a controlled and logical fashion such that demonstration of compliance and repeatable delivery of consistent product is possible.

This statement reflects the company's concern to manage the writing of specifications, and to ensure some kind of control of the activity. Reading the document further, each type of specification is defined under the subheading 'Hierarchy and Rules', with the design specifications ranging from system specifications at the top to software specifications at the bottom, as follows:

- System Specifications
- Product Specifications
- Design Proving Specifications
- Design Requirement Specifications
- Software Specifications.

An examination of the definitions for the first two shows them to be concerned with procedural matters, in that they provide information about contexts for use, rather than their purpose, function, or language form:

System Specifications – Mandatory unless substituted by a Product Specification or Design Requirement Specification.

Product Specifications – Used in place of a System Specification or Design Requirement Specification on simpler and proprietary products. Product Specifications can be used in the selling of products into the open market place where the Company wishes to protect its designs by not disclosing the level of detail normally contained in a Systems or Design Requirement Specification. During development, the product Specification acts as the requirement for all other specifications during which it is at its 'Draft' or 'Preliminary' issue. Once development is complete the product Specification becomes the technical description of the final product and is therefore subservient to the DRS. At this point it becomes 'Approved'.

So far in this document, the information has been essentially regulatory in nature, and it is necessary to read on, to find out anything about language. The section 'Language Style', included in the section on 'Presentation and Format', comprises a total of 162 words on language style that relate mainly to the use of abbreviations, symbols, and graphics. Language aspects are covered by just two sentences, as follows:

6.2.1   Specifications must be phrased in a language free from ambiguity and in such terms that contractual implications are clear and enforceable.

6.2.2   A clear distinction should be given between those statements that are mandatory and contractually binding or non-mandatory and express an aim/recommendation using appropriate wording.

Commenting on these guidelines, one engineer wrote, 'Good advice, but doesn't tell you how to do it!' He was referring to his difficulty with expressing them in clear unambiguous English.

Short though this advice is, it nevertheless reflects two major concerns that engineers raise continually about specifications, that is,

1. Specifications should be clearly and unambiguously written.
2. Specification statements that are mandatory should be clearly distinguishable from those that are not.

5.3   The Customer Requirement

The customer's need usually will have been identified in a document called the Customer Requirement, or, as engineers refer to it more briefly in talk, the Requirement. It is the Requirement that provides the impetus for the writing of proposals and specifications. The Requirement is not an argumentative or persuasive text. It does not have to be, since those responsible for it know that others will compete to provide what is being specified. It is the readers of the Requirement, the engineers involved in the bid, who will need to persuade the writer of the Requirement to accept the proposal they make, and later, who need to ensure that all aspects of the Requirement are catered for in specifications documents. Ambiguity of terminological usage is worth commenting on here: in the case of the Requirement, 'customer' may be an ambiguous term, referring either to

a. the originator and/or writer of the Requirement document or
b. the ultimate user of the product described in it.

The Requirement document is most commonly the trigger for the writing of a proposal, and is the key reference document for specifications writing, if the proposal is successful. On occasion, the Requirement may be expressed orally, rather than in writing, or may begin with a hastily drafted fax or email, an example of which is reproduced in Figure 6.1.

5.3.1 Cardinal point specifications: customer's wishes and ideas

A cardinal point specification is a specification at the highest level, in which the customer, usually a government department, states the key features of a product he wants to be designed and produced. 'Cardinal point' is a naval term referring to the points of a compass, and it is probable that, originally, cardinal point specifications were written by naval customers. Now, the term is used more generally to refer to a document expressing the customer's wish list at the very start of a tendering process, in the 'pre-design' stages (Figure 5.1).

The extract shown in Figure 5.1 is taken from a Requirement, the YGO46 Requirement, in which the customer states exactly what he wants, and is trying to influence the design process by specifying how the products, two medium-calibre gun systems, should function, perform, and be constructed. He is 'calling the shots, so to speak, and saying "this is what it should do, what it should look like, how it should be used", and so on', to quote an engineer's words. Unlike Figure 6.1, which is a brief email, this Requirement is a more substantial, bound document, comprising c.7400 words and 28 pages. It reflects the customer's ideas for a gunfire control system, is detailed, and, indeed, imposes many constraints upon the design engineers, while at the same

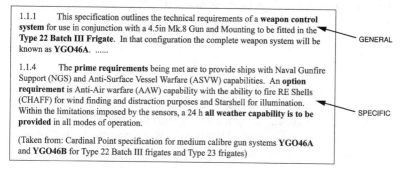

1.1.1   This specification outlines the technical requirements of a **weapon control system** for use in conjunction with a 4.5in Mk.8 Gun and Mounting to be fitted in the **Type 22 Batch III Frigate**. In that configuration the complete weapon system will be known as **YGO46A**. ......   — GENERAL

1.1.4   The **prime requirements** being met are to provide ships with Naval Gunfire Support (NGS) and Anti-Surface Vessel Warfare (ASVW) capabilities. An **option requirement** is Anti-Air warfare (AAW) capability with the ability to fire RE Shells (CHAFF) for wind finding and distraction purposes and Starshell for illumination. Within the limitations imposed by the sensors, a 24 h **all weather capability is to be provided** in all modes of operation.   — SPECIFIC

(Taken from: Cardinal Point specification for medium calibre gun systems **YGO46A** and **YGO46B** for Type 22 Batch III frigates and Type 23 frigates)

Figure 5.1 Extracts from a customer requirement document

time placing the onus on them to create a system which will perform in the way specified.

It is a difficult task, described by one engineer as 'easier said than done'. He added that it may be perfectly possible to fulfil the customer's wish that the system be capable of registering six targets simultaneously and, at the same time, perform two full computations on two of them. However, he went on to explain it was difficult to design a system to these specifications that would also achieve this performance accurately and efficiently at sea, in any weather condition, in any climate, and/or at any latitude, which is what the customer wanted. Naturally, all this had to be done within the customer's budget.

From general to specific

The main purpose of any Requirement is to provide as accurate and precise a description as possible, so that the engineers whose proposal won the bid can produce the system the customer wants. Typically, the customer's description may move from the general to the specific, as shown in the extracts in Figure 5.1. First, in the Introduction are points which refer to the system as a whole, and which provide a context for the lengthy itemised description that follows. Points 1.1.1 and 1.1.4, for example, categorise the YGO46 product (it is a weapon control system), specify where it will be used (on Type 22 Frigates), and specify nomenclature by stating the name it should be given, and how it should be referred to (YGO46). More than 20 years after this Requirement was written, the name 'YGO46' is still used in conversation and written communication, although engineers describe it as being a mature product, well into its service life, and overshadowed by Type 45 fire control technology.

Prime requirements and optional requirements

The overarching YGO46 requirement is subdivided into more specific prime and optional requirements, and as Figure 5.1 shows, Section 1.1.4 specifies:

1. **Two prime requirements,** that is, that ships should be provided with NGS and ASVW capabilities,
   [NGS – naval gunfire support; ASVW – anti-surface vessel warfare]
2. **An optional requirement**, that is, an anti-air warfare capability with the ability to fire RE shells, and
   [RE – radar echo]
3. **An obligatory requirement** for an all-weather capability.

It is significant that the design should be expressed in terms of compulsory and optional features, and a striking feature of design engineers' interpretation of the Requirement is the close attention they pay to this. Their observance is painstaking, as is the care they take to present different categories of requirements in their written response. In effect, the design undergoes a textual deconstruction, and this chapter attempts to explain the rationale behind this approach. Suffice it to say at this stage that the Requirement is the starting point and main reference point for everything the engineers write in proposals and, subsequently, if the proposal should win, the requirements, and contractual specifications.

Specifying how the documentation should be produced

In the YGO46 Requirement, specification spans both structural and stylistic aspects of the proposal to be drawn up. As well as specifying the names and terminology to be used, including associated acronyms (see Chapter 7), the Requirement attempts to influence the structure of ensuing documentation by specifying the details that have to be included about the system as a whole:

General (system) features

1. the technical characteristics required of the equipment, and a list of the equipment that has to be incorporated into the system;
2. the different functions of the system, for example, that the navigation function should provide latitudinal and longitudinal information about the ship's position, or that the gun should be able to register six targets simultaneously;
3. the roles of the system, for example, as a tool in anti-aircraft warfare, or that it should interface with other systems;
4. the ways it should work with other equipment.

It then specifies each part of the system, for example the gun controller's console, fire control equipment, sensor, and optical sights. It states the main function of each (with few exceptions), and then specifies how it should operate, under headings which include, among others:

Features specific to particular parts

1. *Performance* – that is, how accurately the system should operate, and within which ranges, in different sea and manoeuvring conditions.
2. *Environment* – for example, the temperatures at which the system should be able to function.

3. *Technical characteristics of the gun* – design and construction: sixteen separate requirements, some of which are general, and others specific.
4. *Computation of gun orders* – input parameters.

The Requirement also specifies other aspects of the product, which are usually the concern of others with responsibility for technical publications, technical support, and manufacturing:

Supporting material and non-design aspects

1. *Theoretical studies* – a demand for theoretical calculations as support for performance predictions.
2. *ARM* – a section concerned with the post-production phase, when the system is in use. ARM stands for availability, reliability, maintainability.
3. *Documentation* – specifies the documents which should be produced to accompany the system: in all, 17 sets of documents.
4. *Packaging* – how the system should be prepared for delivery.

The customer's modal verb usage

A striking feature of the Requirement is the distinctive use of verb forms, in particular modal verbs. The customer's expectation in the YGO46 Requirement seems clear and brooks no argument, as the following examples seem to demonstrate:

1. is (also) required
2. should be minimised
3. are to provide
4. is/is to be provided/made/displayed
5. should be met
6. are to be detailed
7. will comprise
8. are (subject to)
9. is specified
10. must be/are to be (capable of) calculating/being inputted [*sic*] performing
11. will be applied/indicated/input
12. must be (possible) to register and to perform
13. is to have/be/be made (of)
14. is/are required to offset/compute
15. (It is desirable) to be able to fire

Such verbs help engineers to distinguish between compulsory or optional requirements, and they rely heavily on them when drafting their specification requirements.

5.4 Control of the design process: the need to manage change

5.4.1 A recipe for success (or disaster)

It is all too easy for non-engineers and those from a non-scientific background to fail to understand the importance of specifications and requirements to the well-being of any engineering company. Most companies are perpetually preoccupied with them. They are regarded as difficult to write and a potential source of design problems and concomitant financial outlays, upsetting both the carefully planned work procedures and the engineers who like to follow them. Get them wrong, and the company stands to lose huge sums, credibility, and respect. The burden of responsibility of an exploding space shuttle, a train collision, or an aircraft crash is heavy indeed. Notwithstanding such disasters, it would be no exaggeration to say that badly written specifications have the potential to be catastrophic for some companies.

It seems that in modern technological societies, engineers are expected to think of everything, of every possible exigency, or that they (or their company) may be held accountable in the kind of 'blame culture' that prevails. Whether or not this is a reasonable expectation is open to debate, but the fact remains that engineering companies, and the engineers who work for them on designing and developing products, are deemed responsible for all aspects of the design and function of any product. Engineers seem to accept and shoulder this responsibility as a natural part and parcel of their profession.

Research journal entry: Traceability, even for 'rough' engineering drawings

I quizzed Nick Stanton about what he told me last Wednesday about RESs, or what he jokingly referred to as 'Rough Engineering Sketches'. I asked him what RES really stands for. He had to think for a while and then came up with 'Registered Engineering Sketch'.

Continued

He was on his way home, and I didn't want to delay him, but he seemed interested to talk and so we chatted on. He explained that they were called RESs because they were usually produced to rectify some mistake in engineering design. They used to do their drawings in books that were kept in the stores department, which used to buy these books in batches, each batch made up of books containing pages with unique numbers. Whenever an engineer needed to have a new book, he simply went and took one, knowing that the page numbers would not be repeated elsewhere. So it was safe to say that each RES or design modification could be recorded and easily traced by dint of the reference number on each page. Yet again, traceability was the reason for this practice, although all the drawings are stored electronically now.

Temporal considerations

A large proportion of engineers' work-time is spent on writing specifications and requirements, and although no one has objectively measured the actual amount of time spent on these, it is generally accepted that specifications take up a disproportionate amount of engineers' time. The situation today is little different from that found around the middle of the twentieth century, when Hicks wrote 'no accounting of technical wordage is available today', although, even then, specifications accounted for 'millions of man-hours of writing time' (Hicks 1961: 228). This calculation is mentioned, no doubt, to impress or appal, but it is not contradicted by engineers more recently surveyed, who estimate that they can spend well in excess of 50 per cent of their time writing specifications. However, they are not always sure that it is time well spent because of the problems they encounter when writing them. Specifications and requirements present problems to both companies and the engineers they employ, involving the former in (sometimes huge) unexpected costs. It would be no exaggeration to say that these problems have motivated root and branch reviews of work practices and writing practices in regular recurring cycles through decades, in an attempt to pre-empt problems that might arise. This continual search for reform demands from engineers major investments of time and energy at work.

Also, it has to be said, they are expected to write a specialised type of writing, by following arcane writing conventions.

Text ≈ Product – ergo, text can be engineered

Engineers like to organise the design, development, and making (or manufacture) of their product, and specify how it will be done in a precise and ordered way. They see it as central to their work practice. Since the early specification of their product is in the form of text, it follows that the engineers like to exercise similar controls over the texts they write, and 'engineer' text in the same way that they engineer their products. This attitude is revealed in the following sentences, written to impress a prospective customer with the rigour that is exercised when requirements are written and stored:

> The Requirements Specification for the Medium Grade Generic IMU was analysed and all text containing requirements applicable to software was captured using the RTM tools.
> [RTM – Requirements Traceability Management; IMU – Inertial Measurement Unit]

> These elements of text were then engineered to produce concise and unambiguous statements of requirement suitable for the development process to continue.

Attempt to pre-empt problems

Specifications have been regarded as troublesome for decades, demanding writing that is clear, unambiguous, and accurate (Hicks 1961). The industry generally has seen, over the years, investments in large research projects, with the aim of improving the writing of them. Researchers in the engineering field have been attempting to describe the problems in numerous research papers and articles in attempts to counteract the problems that arise from poorly written specifications (Meyer 1985, James 1997, Chen *et al*. 1998, Riddle and Saeed 1998), and suggest fundamental changes to working and writing procedures. Their work is part of an attempt across the sector to establish rules for writing specifications, in the belief that the problem lies in the vagaries of the English language. Additionally, engineering academics at universities cite the writing skills of engineering students as being a significant factor contributing to poorly written requirements. It seems, however, that these researchers and academics compound the problem by denigrating the English language and suggesting writing tasks which even highly skilled writers would find difficult, if not impossible, to perform satisfactorily.

A major problem perceived by engineers is the management of change. They are concerned to reduce confusion and misunderstandings that arise when changes are made to the design of a product. They also have a concomitant desire to ensure that changes to the design are mirrored by matching changes to text. The handling of these changes has been identified across the engineering sector as a fundamental problem, having a major impact on engineers' working lives. Their inability to control change is costing them dear. Rough guesses by senior engineers produce sums amounting to millions of pounds. Anecdotal evidence abounds, and there is a widespread acknowledgement that changes to requirements lie at the root of the problem. As the situation stands, they are a perpetual major problem to the industry.

Earlier chapters describe engineers' attempts to be precise and procedural, following methodologies as carefully thought out as their engineering calculations. The design and development of a product, be it hardware or software, is seen as a process which can be plotted and planned, and indeed companies commit resources to a regular cycle of self-examination in the search for better work processes and procedures. It is an understood and expected part of engineering work culture. As an example, in a two-year period, one company instigated three initiatives, specifically intended to learn more about change and to manage it, with the ultimate aim of controlling changes made to the design of complex products. First a small research team working on engineering methodologies published a written document on procedures one year. Next, there was the formation of another group, called the Change Team, who worked with their engineer colleagues exclusively on improving work methods. The following year, another group was formed, this time called 'the Task Force', whose brief was, again, to bring about changes in the way engineers worked on design.

These are typical of a perpetual round of engineers' attempts to manage their own work behaviours, accompanied by intense learning of any new software that claims to help them with this. These software packages aim to help engineers with the design, test specification, and test procedures when developing software, although their success in achieving such aims is moot. RTM software, mentioned in Chapter 2, is an example of such software taken up by companies, only to be superseded by another that makes similar claims about enabling engineers to keep track of, and manage, requirements writing.

Research journal entry: How complete can a specification be? No happy resolution possible

More than twenty years after it was written, there arises a problem with the original specification. The story is this. The company has received an urgent request for replacement lenses (belongs in the gun barrel on a ship to help with alignment to/with target), because the ones that were just sent to the Navy didn't fit the brackets. Two ships out there in the middle of an ocean somewhere that need cameras by Monday 28 February. They will claim that the ones that were delivered are not fit for purpose. If they don't get replacements for Monday, it might hit the papers.

Apparently the sub-contractor (Hastings) had changed the lens supplier without telling the company, so that although the lens performs all the functions, it is a different size from the previous lens (2 mm too long), and therefore useless. Aaron discovered that the lens was the wrong shape when he tried to fit them into the bracket fitting. He argues that had Hastings done the same thing, i.e. fitted them into the fittings, they would obviously have discovered that for themselves. Hastings fobbed Aaron off all day yesterday, and I witnessed four or five engineers at times standing around (wringing their hands metaphorically speaking) as he spoke on the phone. Today, gloom has descended over the area because there seems to be no resolution, and Aaron has to tell the Navy that there is no solution to the problem. It seems the original specification specifies the actual make of lens, and Hastings had recently changed the make, without telling them. However, unbeknownst to anyone, in the small print of the contract Hastings states that it reserves the right to change suppliers. So there have been huge arguments about the legality of changing the supplier without ensuring the lens not only delivered identical performance, but also had the same dimensions.

This problem points to two causes: the original specification was incomplete and based on a relationship of trust (i.e. that the sub-contactor would see to all exigencies). The two are interrelated. In the end, with the trust broken, acrimony sets in, and with it, the need for retribution and accountability. The Navy will get even angrier when it discovers its guns can't fire properly because the recently delivered lenses don't meet the original requirement, and there will almost certainly arise the need to allocate blame.

Continued

Engineers may come and go, but ultimately, it is the documentation which endures. All parties will refer to the original specifications produced here by this company in the event of an enquiry, or the need to compensate through legal channels. In fact, this reference to the original documentation has already happened, and these engineers have discovered:

1. that the requirement/specification was a draft, i.e., had no official signatories. The significance of this may need to be determined. So far as the recipient is concerned, I don't see that this should diminish the strength of the document in legal terms. Hastings would treat it as they would treat any requirement, surely.

The company, on the other hand, may wish to discover why a final version of the specification was never produced. Why didn't someone take responsibility for seeing this was done? General opinion seems to be that those concerned were simply too busy, and it was overlooked. It was, after all, a 'very small job' from the engineers' point of view, worth c.£12 000 ('no reason to get out of bed for that kind of money...they're so difficult this kind of contract, so small and special' [Jim])

2. the specification was incomplete. It is insufficient to specify a type of brand for a part of the whole. Clearly, the manufacturer may go out of business, or discontinue the line and stop producing the part. NB: So we could say that for ALL projects a good technical description will describe the function, dimensions, and material (and/or mode).

A final observation: at such times of dispute, engineers at the company, and probably at Hastings too, inevitably turn their attention to who is at fault. It's stressful for the engineers at the time and distracts them from focusing on the customer. Ultimately, they are prevented from working as efficiently as they would like. This may be all part of the job, but wouldn't this all have been avoided if decent specifications had been written in the first place? Whatever, Aaron has been trying to avoid an escalation of the row, especially as he needs to ensure a steady supply of new work. So treading a fine line.

5.5 Categories of specification

Strictly speaking, 'requirements' are hypotactically related to 'specifications', which is a superordinate term, so that specifications include requirements. Requirements, in turn, may occasionally be referred to in contrastive terms as dichotomies: they are hard or soft, high-level or low-level, and formal or informal, open or closed, restricted or unrestricted, depending on the context. Little discussion about these terms exists in the literature, and clear definitions of them are lacking. This dearth of information may be due to their relatively infrequent occurrence in writing. Engineers tend to use them more frequently in spontaneous discussion, where it is taken for granted that everyone knows what they mean. The problem is, though, that they do not know precisely, which is curious, considering the precision that is demanded from engineers when they write the specifications.

5.5.1 Hierarchies in requirements

When combined together, specifications and requirements are seen as defining the design task that engineers need to complete. So far as this study is concerned, it is useful to recognise that requirements (from now on any discussion of 'requirements' also applies to specifications, unless indicated otherwise) are hierarchically determined, there being three main levels, although there may also be intermediate levels of requirements.

High-level requirements expressed in a document are sometimes referred to as the Requirement although they may be more formally referred to as cardinal point specifications, or Invitation to Tender (ITT). High-level requirements describe what the customer wants in more general terms, highlighting the key features of the product that the customer wants, described earlier in this chapter.

The concern amongst engineers involved in research and management and those in academe about the impact of poorly written specifications at the higher levels has already been mentioned. It has long been observed that high-level requirements that are not clearly expressed often ill-affect requirements written at the lower levels. It is a reflection of design engineers' aspirations that they continue to search for, or to invent, a system to translate ordinary, 'natural' English language into lower-level requirements or code. It is, in effect, a search for yet another holy grail, except that this one has proved particularly elusive. Requirements are a specialised kind of technical description. 'Description', as the previous chapter explains, is both a useful and yet inadequate word to account for the writing that engineers have to produce. Their descriptions differ from traditional depictions which tend to list description as a type of writing alongside other types, like argumentation, narration, and exposition (Brooks and Warren 1952). To read explanations more relevant to the complex nature of engineers' descriptions, it is necessary to study the work of those with an understanding of scientific and technical language (Marder 1960, Weisman 1962, Trimble 1985, Dobrin 1989, Halliday 2004). In general terms, 'description' is but one communicative function performed in the engineering workplace. More specifically, it is possible to discern different categories of description and to locate them on a technical description continuum according to the level of detail being described. In the area of design, for example, engineers refer to 'high-level' and 'low-level' description. Such description can be seen in terms of a hierarchy, which, broadly speaking, comprises description types that are more or less general (or specific). The broad, albeit over-simplified, picture is shown in Figure 5.2.

Low-level description may alternatively be referred to as 'sub-system' description, and specifies the finer design details of the product. This is where the use of modals, discussed later, becomes sensitive. Engineers commonly have difficulties with producing detailed, 'low-level' descriptions to reflect higher-level description, particularly when writing software code. Ideally, the writing process should have a 'waterfall' effect, with the higher-level concepts 'trickling down', or 'cascading', to use business jargon. The idea is that understanding of the design is assumed to cascade down the 'levels' of design so that engineers working at different aspects of the design interpret it to create the necessary machine codes at the 'lower' levels.

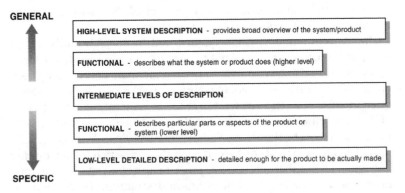

Figure 5.2   A hierarchy of technical specification

However, when changes are made to high-level descriptions of a product, problems occur when lower-level requirements have to be re-written as a consequence of the high-level change. In a discussion about making changes to requirements, an engineer explained it thus:

> People wonder why there's all this fuss about changing a single sentence of high-level description. It's because that sentence may give rise to 500 new sentences of code [at a lower level]. And I think sometimes engineers make a [low-level] change and find it too tedious to record why. And then when they have to return to it later on, to make another adjustment, the engineer – or it may be another engineer – may find that what was blindingly obvious at the time, becomes opaque, and he can't fathom why on earth he made the change in the first place.

So, actual practice shows that communication does not always flow smoothly between the levels, and there may be a breakdown in this process, especially if engineers writing the code are not involved in design decisions made by the systems engineers, the high-level designers. The (low-level) engineer cannot really see how his particular bit of coding fits into the wider design concept. Also, it sometimes happens that a 'high-level' design concept needs to be revisited and modified in the light of results being produced at the lower levels, but breakdowns in communication mean this is not always communicated effectively between engineers working at different levels of design description. It is understandable, then, that mismatches between descriptions at different levels result, and that the requirements may be deemed inadequate.

5.6 A special language: engineers have devised their own rules

5.6.1 What engineers don't like about 'English'

'Thou shalt not be vague'

So far as specification writing is concerned, it seems that engineers believe English to be inadequate for their purposes. This impression has been gained from engineers' opinions garnered over two decades, and from their reading material, which includes textbooks, engineering journal articles, and writing guidelines. A typical opinion is that

'English' encourages engineers to be ambiguous, incomplete, and incorrect. Engineers think it is open to interpretation and that this makes it difficult for conditions to be precisely defined. A consequence of this is that two people reading the same statement could come up with different interpretations, and this could be a potentially difficult and expensive situation for the company when the misunderstanding occurs between the customer and the design engineers. Such mismatches between customer expectation and engineer provision happen more often than engineers would like and, furthermore, are encountered across the engineering sector on an international scale.

One engineer explained the problem in this way:

> The vast majority of problems arise from misunderstanding what your requirement is supposed to be. Quite often, you're an awful long way down the design process before you actually find you don't understand what's going on, or that you misinterpreted something... It's the ambiguity and incompleteness. People misinterpret what was written. The customer has one set of domain knowledge, the engineer has another. There is an intersection between the two, but because there is a difference in their knowledge, each may interpret a given statement in a given way, both of which are correct, but you may end up implementing something different from what the customer intended.

There also seems to be the impression that natural English leads writers to make errors when drafting requirements, because it encourages the construction of illogical expressions, as these engineers explain:

> People can make an awful lot of mistakes in English: mistakes in logic. English will let you express a lot of illogicalities, and to an engineer that's impossible.... I think English is the problem. It's a language to express feelings and emotions; it's a language with shades of meaning. It's not a logical thing.

> I'd like to see a situation where no English was used at all. I'm not interested in problems with English anymore, and want to banish it altogether.

Put simply, there is an assumption among the engineering fraternity that everyday English is unsuitable for their needs. They usually call this kind of English *natural* English or *informal* English, and have made a distinction between what they call *natural* or *informal* English and *formal* English. This attitude towards English in engineering industries is equally matched by concern amongst engineers in academia, who are concerned about how to improve the clarity of engineers' writing. Academic journal articles and text books make references to the inadequacies of English, as exemplified by the following, the first from an article on how specifications should be phrased, and the second from a textbook:

> Req.I12 is interesting because it illustrates the deficiencies of English. (Fitzgerald 1993)

> Unfortunately, natural language has been found to be of limited use in the production of precise specifications. Natural languages have a variety and richness that tend to militate against precision: ambiguity and misinterpretation abound in natural language descriptions. This is not to say that natural languages cannot be used for specification purposes; it is just that to gain the required precision leads to documents that are so long that their sheer length becomes a problem. (Thomas *et al.* 1991)

In their search for the holy grail of unambiguous specifications, engineers have tried to make up for the short-comings, as they see it, of 'natural' English by devising their own language rules. Their attempts to do this, by systematising the use of modal verbs, have added an extra layer of complexity to a writing task which begs to be understood and simplified.

5.6.2   The special case of modal verbs

Modal verbs can convey very subtle shades of meaning in English, but have proved enduringly problematic. In trying to exercise controls over writing, engineers' attempts to use modal verbs, particularly the modal auxiliaries 'will' and 'shall', create confusion (Sales 2000). A kind of engineering grammar exists in the engineering community, which advocates that 'will' should be used in statements to express a feature that may be desirable in a product, but is not compulsory. 'Shall', on the other hand, has some kind of mandatory force when used in third-person constructions, for example,

> If power is removed from the unit while the system is held in the Power-up State following the detection of a fault, the system

shall unconditionally leave the Power-up State in order to enter the Off-State.

This is clearly a crucial mandatory feature, emphasised by the engineers' use of the adverb 'unconditionally' in 'shall unconditionally leave the power-up state'. It is interesting to note, though considering the mandatory force of 'shall', that the authors of this requirement found it necessary to add weight to it with 'unconditionally'. Clearly, the engineers considered a mere 'shall' to be inadequate in this particular case.

Kirkman discusses problems arising out of the use of these particular modals, urging engineers to use other less confusing ones instead (1992: 124). However, as will be seen later, 'will' and 'shall' are firmly entrenched in the engineering lexicon, with engineers steadfastly claiming they convey these special meanings.

It is possible to see a link between grammatical categories of mood and the writing of engineering specifications, because specifications are primarily concerned with interpreting the customer's needs and desires in terms of requirements. In fact, when talking about the customer requirements, engineers frequently refer to them as the 'customer's wishlist'. The customer's statements of need clearly convey basic modalities, as shown in the following two examples taken from a customer's original requirements specification:

1. A below decks operating position *is to be provided* in the Control Room.
2. *It is desirable that* the system is hardened against the effects of an explosion.

In these examples, the customer differentiates one from the other through the use of different verb groups. He clearly wants the first and, by writing 'is to be provided', makes it plain that there is no choice in the matter: a below decks operating position must be provided in the Control Room. There is room for manoeuvre with the second, however, and this is signalled by the use of 'it is desirable that': he would like the system to be hardened against the effects of an explosion, if it can be done, but if it cannot be, and the engineers find it impossible to achieve, he will not reject what they come up with.

The significance of 'shall'

When interpreting the customer's Requirement, engineers emphasise the need to use particular modal verbs correctly, especially 'shall' and 'will'. Without exception, they stress the importance of 'shall', because they have seen all too often that inaccurate use of modal verbs has led to misunderstandings and disputes with the customer. There is no doubt that the use of 'shall' in a specification indic- ates an action which must be done, for example (my underlining),

5.7.6.3.1.1   The system <u>shall</u> accept positional data from a closed loop servo system which places the system in both elev- ation and training in response to demand signals....

5.7.6.3.2.1   The daylight TV camera <u>shall</u> convert optical images into composite video signals for TV monitor display and video recording purposes.

5.7.6.3.2.2   The THIM <u>shall</u> convert Infra-Red (IR) radiation into composite video signals for Video monitor display and recording purposes.

Those who ask why the use of 'must' does not suffice, soon learn that the use of 'shall' has the function of signalling a requirement which must be carried out, and which is mandatory in contractual terms; it is a legal requirement, in other words. Any engineer will provide a quick gloss on this particular usage by explaining that a 'shall' indicates what must be provided for the customer, whereas a 'will' indicates what is desirable. When pressed, most agree that desirable attributes of a product may be expressed with structures other than 'will', a fact borne out by even a cursory examination of specification documents, but opinion is unan- imous about the mandatory force of 'shall'. There is an unmistakable attitude of respect towards any use of 'shall' in specifications; the word seems to be accorded the highest status indeed so far as engineers are concerned.

This rule regarding 'shall' and 'will' is not restricted to the United Kingdom. I recall a conversation in Singapore with two academics, a Dutch and an American. Both railed against this peculiar usage there, describing it as illogical and ungrammatical, and blaming it on reac- tionary British engineers who had imposed their entrenched practices on former British colonies, and, indeed, at the time their argument sounded plausible. It is useful, however, to take a broader, global view: this usage is international, and considered the norm across the globe. For example, German engineers are well used to working with British or American counterparts, who talk exactly the same language so far as 'shall' and 'will' are concerned. Advice about the status of these modal verbs is readily available in engineering handbooks, text books and the like, as shown in the following example which is an extract from a writing manual issued by a certification body (ITSEC, discussed in 2.2.3):

> Within the criteria certain verbs are also used in a special way. <u>Shall</u> is used to express criteria which must be satisfied; *may* is used to express criteria which are not mandatory; and <u>will</u> is used to express actions to take place in the future. (ITSEC 1992: Para 0.12)

Another example, this time from a text book (Texel and Williams 1997), reiterates the obligatory force of 'shall', but also highlights another function, significant for engineers using specialist software to monitor requirements writing: that it is used as a marker in computer databases, so that engineers can easily trace all the 'shall' requirements to see if they have been implemented.

Definitions of 'shall' and 'will'

1.2 Capture 'shall' statements

Purpose

The purpose of this Activity is to produce an initial Requirements Trace Matrix (RTM) that contains the entire set of sentences from the System Specification, and any other agreed-upon documents, that include the word 'shall'. A sentence that includes the word 'shall' represents a requirement that must be satisfied. . . .

Definition(s)

'**shall**' **statement**: A single '**shall**' **statement** is a sentence that includes the word '**shall**'. A '**shall**' **statement** is extracted from the System Specification (and any other agreed-to documentation). A '**shall**' **statement** indicates a contractual requirement for the system to be developed. (Texel and Williams 1997: 22)

5.6.3   Monitoring requirements in computer databases

'Shall's versus 'will's and the others

RTM, and other software like it, is used by engineers to monitor the progress of their design work as they produce increasing amounts of written product description. They need to keep track of all the detailed requirements that have been specified in the requirements specification, which itself is a reflection of the system (or high-level) specification. The previous section explains how 'shall' requirements are 'mandatory' and have to be carried out (or 'implemented', in engineer-speak). In order to find out how many 'shall' requirements have already been implemented, RTM has a way of checking the requirements to see if they have been recorded as 'met'. The tool enables the engineer to comb through the requirements, sometimes thousands of them, identifying all the 'shall'-ones, and then tagging them. These are then matched with the high-level system requirements, and when they have been tested and 'approved', these tasks can be 'ticked off', as it were, as having been completed (or implemented).

The 'shall' requirements which have not been implemented are still outstanding, and have to be worked on so that they too can be 'ticked-off', in order for the system to be compliant. A snap-shot survey of one project, for example, found there to be a total of 430 requirements, of which 327 were 'shall' requirements. The remaining 103 contained other modal verbs, which were non-mandatory. Only 27 of the 'shall' requirements (just over 8 per cent) had been implemented, whereas 52 of the rest (50.5 per cent) had been. The investigating engineer suggested it would be interesting to examine why significantly more non-'shall' requirements had been implemented, compared with the 'shall'-ones, which had not. The findings seem to indicate that requirements in the 'shall' category are more problematic for engineers.

It is apparent, then, that the design process is not nearly so neat and tidy as engineers would like it to be, nor the customer as controllable. The dynamics of the process mean that aspects of the requirements change, which in turn affects the design, which has to be re-written in order to match these changes.

Inconsistencies of practice

Specification documents seem to indicate that engineers do not practise what they preach, at least not consistently. There are occasions when they go against their own prescriptions and use modal verbs in a variety of ways, within and between documents. A survey of 'shall' usage in four specification documents, two for hardware products and two for software, reveals the following:

1. A section comprising 65 requirements yields not a single *shall* (hardware requirements). The section contains modalised structures, many of which could convey mandatory obligation, however, as shown in the following:

   • *must*, as in the following example: 'Appropriate software <u>must</u> be provided with the system to allow operator performance assessment.'

- *will*, as in 'For the System, the KLK Mk5 version of the Pedestal Sight will be used.'
- *is/are to* + *lexical verb* structures, as in 'The purpose of the GLKV is to provide fall-back modes of operation.'

2. A relatively short requirements specification uses 'shall' in 30 out of a total of 42 individual requirements (hardware requirements).
3. A longer document includes 'shall' in each of its 2255 requirements (software requirements).
4. Another comprises 280 requirements, all of which are 'shall' statements, bar five (software requirements).

These differences are, in part, a reflection of differences between software and hardware requirements and how they are composed. The former are drafted exclusively with the aid of electronic writing tools, which themselves are devised specifically to control engineers' language. As a result, software requirements, among other things, reveal a degree of uniformity and conformity with the conventions governing the use of modals, so that every requirement contains a 'shall' modal verb phrase, usually without exception. Nevertheless, and in spite of these specialist tools, software engineers report no reduction in misunderstandings that arise with the customer, and also between themselves.

In contrast, hardware engineers' requirements tend to be more variable, exhibiting markedly less conformity with rules governing modal usage. Hardware engineers may choose not to use computerised writing tools, and so are less constrained when composing requirements. They freely admit they do not use modals in the way prescribed, but are nevertheless concerned about expressing the requirements unambiguously and clearly. The following, for example, is a mandatory requirement:

> The system **is to be** capable of calculating minimum range for crest clearance...

which ought to have been expressed thus:

> The system **shall be** capable of calculating minimum range for crest clearance...

At issue in the above example is a purely mechanical matter of substituting 'shall' for 'is to be', although numerous other examples exist in companies. Engineers do not like to break the rules, but the fact is that they do, and they do it often, even though unintentionally.

Explaining why engineers seem to disregard their own writing conventions is beyond the remit of this book, but discovering the reasons for this behaviour would make an interesting project indeed.

5.6.4 A specification document case study

A specification document is now examined to gain a clearer impression of modal verb usage. This document, which is fairly representative of its type, provides a set of hardware requirements for a portable computer component, and the results of this study seem to confirm earlier observations that modal usage in requirements is inconsistent. The inconsistencies seem to be fairly typical, epitomising the characteristics of hardware requirements documents examined so far.

At the very beginning, the document includes in its introduction a list of definitions, which most would agree is good practice, that is, to define how terms are used. However, these contradict published guidelines, in particular those regarding 'will' and 'shall', which are deemed in this document as having the same mandatory force. Prescribed usage for these auxiliaries, along with a list of abbreviations and references, is reproduced verbatim below:

DEFINITIONS
Within this document, the following terms shall be interpreted as. described
'May'   Allowable.
'Might'   Allowable.
'Shall'   Obligatory (except where mentioned in a note).
'Should'   Preferable.
'Will'   Obligatory.
'Would'   Preferable.

Although 'would' and 'might' are included in the definitions list, being defined as 'Preferable' and 'Allowable' respectively, they never actually appear in the requirements themselves. 'Shall' and 'will', however, occur most frequently in the document: 'shall' in 44 of the 97 requirements statements, and 'will' in 32. On the other hand, certain modals have been excluded: 'must' and 'can' are actually used within the document to express a small number of requirements, but are missing from the list of definitions.

Furthermore, modals are not used as prescribed: either modals designated as having mandatory force (a) are used to convey other (non-obligatory) meanings or (b) are absent from structures intended by the engineers to be mandatory. The latter is the case in eight of the requirements, of which the following is an example:

> The DTM (including cable and PSU) is required to be a portable item.

[It seems that engineers' problems with specification writing are not restricted to modal verb usage and interpretation: engineers agree that this is a mandatory requirement, but disagree over the interpretation of 'portable' in design terms.]

Conversely, the 'obligatory'-designated 'will' is used, as shown in the following requirement, but not to convey mandatory force in forbidding the action; instead, it describes future intention and prediction in the following examples, respectively:

> Due to the simplicity of the DTM, reliability analysis <u>will not be undertaken</u>.
> Authorisation of further development or manufacturing release <u>will depend</u> on successful completion of formal reviews.

A pattern of usage is emerging indicating that mandatory force is also conveyed in modalised structures surrounding 'be', as evidenced by 13 of the verb phrases in the document: a significant proportion of these (8) follow the pattern: modal auxiliary + 'be'[+ adj.] + prep. + complex NP, as in the following examples:

1. shall be subject to the procedure set out in Ref. X;
2. will be limited to 3 man-days;
3. shall be valid from the delivery date to ABC plc;
4. should be of sufficient length to allow the DTU to be removed from the case for changing.

The document contains another nine requirements intended by the engineers to be obligatory which do not contain an 'obligatory' modal from the definitions list. Some are catenative constructions, for example 'are expected to be', 'are expected to discard', 'is to be produced', 'is required to be', as exemplified by the following:

> A Master Record Index (MRI), showing the design standard of the equipment is to be produced in accordance with ISO 9001 approved Supplier's procedures

Furthermore, there are two non-modal structures with mandatory force:

1. In the latter case, the connections required are as follows:...
   [followed by data tables].
2. The Supplier is responsible for carrying out these tests.

5.7 Concluding observations

It is useful at this stage to return to the bespoke tailoring analogy mentioned in Chapter 1 and expand on it, as this may help us to view engineers' labours over writing specifications in perspective. Design engineers may be likened to bespoke tailors. In so being, they are subject to the vagaries (even fickleness) of the customer's wishes and needs. The customer is only human after all, and like the man looking at himself in the mirror in his new tailor-made suit, the tailor hovering with measuring tape in hand, ideas come to him about how to improve the garment to make himself even smarter, more comfortable, more elegant, and so on. When engineers present textual 'mock-ups' of a nearly completed design, this seems to have a similar effect on the customer, who finds new ideas and possibilities about the design occurring to him that he had not thought of before. The engineers wish to please the customer (they are always inclined to hone a design anyway), and so they embark on making changes accordingly. These may inspire other new ideas that occur to them or the customer, with the result that the chain reaction of design change is perpetuated. Such is the way of any creative process.

Attempts to exercise tighter controls over tendering procedures, the procurement process, and engineering design processes will continue, but engineers' ingenuity and inherent creative nature will stymie these (no doubt worthy) attempts from time to time. It would therefore be misleading to state that engineers' writing skills are not up to the job of drafting specifications, just as it is an over-generalisation to claim that engineers are not good writers. Such assertions only muddy the waters of what is clearly a complex writing situation. This is not helped by the lack of clear guidelines regarding the use of modal auxiliary verbs, for example 'will', 'shall', 'may', 'might', 'would', 'could', among others. Existing guidelines serve merely to sow confusion among the engineer writers, leading to inconsistent usage and difficulties of interpretation. Any advice that is available on using these verbs, particularly 'shall' and 'will', is conflicting and often leads to misinterpretations of these modal meanings in the context of specifications between engineers and their customers, and between engineers themselves.

6
The Bid Process and Persuasion

6.1 The bid process

This chapter examines the background to proposal writing, describing it from the viewpoint of engineers who prepare the documentation. They would be members of a bid team whose main concern is to persuade the customer that theirs is the best solution for his needs. An essentially human perspective is provided of significant events to show what happens when a bid is being prepared. These involve the writers and readers of proposals, the approach engineers take to writing persuasive text, and the impact of winning or losing in a competitive-bid situation. The chapter also describes what the customer needs to be persuaded about and how the engineers try to achieve this, in spite of their unease over using persuasive language. This discussion is intended to provide a context for considering later chapters that describe the presentation and structure of technical proposals and executive summaries.

The information in the chapter is divided into two sections. The first explains the bid process, telling stories from the workplace that show its effects on those taking part. Material for this section is drawn from journal entries to a greater extent than in other chapters to provide textual 'snap-shots', describing the unfolding of events, big and small, in the bid process. As a result, the first section touches on issues relating to human and ethical dimensions that have an impact on text production. The second section explores the notion of persuasion, which engineers feel ambivalent about. It describes their views about language in persuasive texts, and the strategies they follow to be persuasive in a surreptitious sort of way.

6.1.1 Proposal writing consumes huge financial resources

The actual consumption of financial resources across the sector by proposal writing, and the concentration of engineers' creativity, time, and effort on proposal writing, is massive. It will be seen that engineers view proposals with some ambivalence, since they are uncertain whether or not they are wasting their time on them. They have to apply themselves to the task without knowing if it has been worthwhile until weeks or even years later.

Engineering companies, particularly those in 'western' economies, are experiencing a prolonged, difficult period, and as a corollary of this, proposal writing has become one of the most important writing activities in the last decade for those companies looking to remain competitive. Engineers have been caught up in bid activity to an extent unforeseen a decade or so ago. They are an integral part of bids for more business, since they must describe their designs for products to potential customers. This activity has significantly broadened their writing horizons and the range of communication tasks they perform.

In a fast changing market, companies have tried to reduce reliance on government contracts by diversifying. Engineers have been affected by this marketing shift, responding to the need to bid for more civil contracts. However, calls for a change in direction on the part of some managements, who would like to secure more civil/commercial contracts, have not changed the traditional relationship with governments that endures, especially between aerospace engineering firms and government defence departments.

The last decade has seen a channelling of engineers' efforts into writing proposals. A simple indication of this is the larger proportion of company training budgets being allocated to commissioning courses in proposal writing for their engineers. Also, engineers themselves claim a greater proportion of their time is now spent on proposal writing. This is certainly a new departure for them, forcing a change to their traditional writing habits, and making writing demands of a different kind upon them. It has also brought about a sea-change in their attitudes to writing, since proposal writing makes rhetorical demands that they used to regard as peripheral to their work, for example, persuasive texts, side by side with the technical description they relate to more closely. Another causal factor in this attitudinal change is team working and team writing, which demands of engineers a broader perspective to their writing. This means, in practical terms, that engineers who are responsible for preparing a bid need to take into account all aspects of the product they are offering, including not only the design, wherein lies their main interest (some would say 'passion'), but also other aspects like manufacturing, servicing, and contractual legalities.

Research journal entry: Proposal writing is an expensive activity and a risky business

Proposal writing seems to be made up of bursts of expensive activity. Malcolm mentioned how a proposal can involve quite a few people, as for example, in Project Sunrise. He said there have been as many as 20 people at the company involved in it at any one time, over a prolonged period and at a cost of hundreds of thousands of pounds. He told me to imagine similar teams in other companies across the country, and across other countries [Italy, France, etc.]. The cost of the project to these companies must run to millions of pounds. To get an idea of the sheer scale of it all, each company worked on a tiny bit of the ship, each part coming together to form a whole ship. Malcolm's team was designing the thermal imager for it.

At quite an advanced stage the politicians decided to pull the project, leaving all the companies with nearly finished proposals but no one to propose to! Purely for political reasons it seems. The rumour goes that British politicians suspect that the Italians, who said they would buy c.6 ships, would back out weasel of the deal and buy only two; and that the French were playing the same game. Britain was committed to buying 12 and was fearful of being lumbered. That's the gossip anyway. Malcolm made the point that these companies have to get the money back somehow, and that governments probably pay high prices to keep engineers through 'dead' periods, since the companies recoup their losses when they win proposals. I'm not sure how the economics of that works out (and what about the companies that don't win? 'They go under', said someone, 'look at Ferranti, Marconi...') but it made some sense at the time we were speaking.

6.1.2   Types of proposal

Anyone who has composed a document with the purpose of bidding for funds or competing for business will understand the intent underlying engineering proposals. An engineering proposal is a formal and complex document, written by a team of engineers, together with their commercial and marketing colleagues in highly confidential working conditions, as part of a tendering process. Usually, the proposal is competitive, and submitted to the customer with the aim of being short-listed and, ultimately, selected as the winning proposal. Occasionally, the proposal may be non-competitive, when the customer has a need for a product or service and asks for suggestions from the company.

Design engineers have to write two key parts: the executive summary and the technical section of the proposal. If the two are submitted together, they are referred to as a single textual entity and called 'the proposal' or 'the technical proposal'. For smaller bids, the technical proposal may be submitted as a single complete proposal. At other times, the technical proposal may be a sub-section of the whole proposal, as is shown in the next chapter, which lists the different sections to be found in larger proposal documents. So, the engineering (or technical) proposal may be submitted as a technical volume, to be one of a tripartite set of proposals, the other two parts of which are usually prepared by the commercial and legal departments of the company.

The RFIs are examples of other documents produced in response to queries from potential customers. Engineers commonly refer to such responses as RFIs, although more logically they should be called RRFIs, that is, 'Responses to Requests For Information'. RFIs are usually shorter documents or may be in the form of a letter, but are written in the knowledge that they could lead to future business for the company if the potential customer is impressed enough, and persuaded, to invite a formal proposal as a result of reading it.

The bid team writes any proposal with the aim of persuading the customer to place it on a short-list and, ultimately, to be the one selected for the prize, which, in this case, could be the winning of a business contract for the company. The 'business', so far as the engineer is concerned, relies upon the design, production, and delivery of an engineered product, which, put simply, could be hardware or software, or a combination of both. Proposals can be large or small, ranging from those with the potential to earn tens of thousands, for example 'rehosting' an existing software product into a new aircraft, to those worth hundreds of millions of pounds, as in the more recent case of, for example, the Joint Strike Fighter.

Research journal entry: The need for extreme secrecy sometimes

Phil pointed out that there are project documents whose titles may not be classified, but whose contents are, and projects whose very existence is secret and unmentionable. In the case of the current bid, it is important not to release information that the company is a bidder, because even those companies submitting bids need to remain incognito and unknown to each other. Reduces the likelihood of corrupt practice, of course, or at least it's an attempt to.

6.1.3 Readers and writers of engineering proposals

Prospective readers and the executive summary

These are the target readers for the proposal, who are the people representing the customer. They may be a government department or another large company, and can be seen to comprise two main categories:

1. *chief readers*, who make decisions about proposal selection, and who usually read only the executive summary. Chief readers may not be engineers, but are usually the chief executive or a senior member of the management team.
2. *team readers*, who scrutinise and vet parts of the proposal. In the case of the technical sections of the proposal, these are usually engineers with the same professional expertise as the proposal writers. Awareness of this fact, that the readers are engineers like themselves, can make the engineer writers rather more circumspect, as is discussed later in the chapter. The readers work in reading groups or teams, each team representing a specialist aspect of the project, for example engineering, finance, and law. Engineers are aware that these specialist readers probably only read discrete parts of the proposal, and so use the executive summary as a kind of orientational text. It is sometimes copied and presented at the beginning of each volume of a large proposal. In this way, the summary helps readers, who only read particular sections, to better consider the detail of the technical design, say, within the context of the overall proposed solution. Thus, effective executive summaries serve a metalinguistic function by providing readers with an overview of the whole project, enabling them to rise above the detail and to see 'the wood for the trees'.

The executive summary also has the potential for priming readers, by cultivating a particular mindset and predisposing them to the benefits (selling points) of the proposed solution. The readers of proposals are certainly no fools, but some proposals are exceedingly long and detailed documents and, by highlighting the most significant benefits, the summary could persuade a reader to appreciate positive aspects more and to pay less attention to the drawbacks of the solution being proposed.

Chief readers

These teams of readers are usually headed by a smaller group or a single person, who may be a team leader or senior executive. For practical purposes, these will be referred to as chief readers. They directly represent the customer, and it is with them that the responsibility lies to select proposals for the short-list, and ultimately to choose the winning bid. Clearly, it is crucial for the proposal to clear the first hurdle of the bid process and succeed in being selected for the short-list. Chief readers may not always have engineering or other technical qualifications, and are just as likely to have commercial and/or procurement expertise. It is unusual for them to read any other document apart from the executive summary. They rely on the views of specialist team readers to help them decide about matters beyond their expertise and understanding. During the reading process, the readers rate the proposals, using the executive summaries mainly, according to particular selection criteria, and decide which proposals should be rejected or retained for further consideration. It can be seen, therefore, that the executive summary plays a key role in the bid selection process.

Retrospective reading, of a kind

If the proposal fails, the engineers experience feelings of dejection, but quickly dust themselves off, metaphorically speaking, shrug their shoulders, and resume work on other projects (i.e., if keeping their jobs is not dependent on winning). If the proposal wins, however, the design process enters another phase during which the engineers have to revisit the technical proposal. They need to review what was promised and to specify in more detail how it should be designed, produced, and used. At least, this is the theory. In practice, some engineers claim design is always 'done from new' for the following reasons:

1. The Customer Requirement may have been a draft document, in which case a 'firm' Requirement would need to be produced, and then responded to by the design engineers.

2. If the Requirement was firm, a new design team may be brought in to interpret it afresh, or, as an engineer put it, 'for real'.
3. By the time the company learns it has won the bid, nine months, or even one or two years, may have passed, by which time new technological developments may demand a rethink of the product.
4. During the negotiation period, the scope of the work usually changes to match the agreed price.

Research journal entry: After a failed bid, some design remains secret

SKIPE, which was so hot only a month ago, is now considered absolutely dead. They lost the bid, and there was a lot of angst and delving on the part of Andrew and his team to find out what went wrong. The proposal was a large one, and an abridged version of the document is available on the database now. It was a proposal for an anti-missile system, apparently, and the very secret technical bit is stored elsewhere in some inaccessible place. It was written by engineers working in a locked room which could only be entered by a favoured few. It is kept secret because the essence of the solution is based on their calculations about how the attacking missile would behave in different conditions. Anyway, they think that some good work went into it, and they may need to draw on it another time. In the meantime, got to keep it safe and secure.

6.1.4 The writers and bid preparation

The type of proposal and scale of the solution determines who should be involved in writing it. For example, a relatively small non-competitive proposal, worth tens of thousands of dollars, say, may involve two or three engineers, led by the one responsible for the particular product or country involved. He would work with the assistance of a technical author, and written inputs from one or two other specialist engineers. On the other hand, a larger competitive bid, worth tens of millions of dollars, would involve a project team comprising ten engineers, or more, who would work on the overall engineering solution and on writing parts of the technical proposal with a team of technical authors. Engineers usually write particular sections, which are then passed to others for comments. Their colleagues in the commercial and legal departments would produce other sections of the proposal. The process can be lengthy and complicated, with teams usually following an agreed procedure to manage a process that can so easily become unmanageable and get out of hand. However, it is common for bid teams to have little time to prepare the proposal. Sometimes they must have it ready in two months, or even less, working frenziedly to make the deadline. Popular with bid teams is the advice provided by proposal writing consultants and their handbooks epitomised by the likes of Newman (2003), which provide detailed information about managing the bid writing process and document preparation.

At particular points during the bid preparation the documentation is scrutinised by specially convened teams, comprising colleagues not directly involved in the bid. During meetings with the proposal writers, the proposal documents are picked over in detail, and commented on (Newman 2003: 162). In this way, a larger circle of engineers may influence proposal preparation in some way or other, although fewer are directly involved in the writing of proposals, and even fewer still in writing the executive summary, as revealed by a survey at one company, which revealed that 47 per cent of a pool of 200 engineers were involved in writing proposals, and 15 per cent in writing executive summaries.

Larger writing teams generally comprise a preponderance of two types of engineers: those concerned with engineering design and those concerned with after-sales support and maintenance. Those concerned with management and commercial aspects of the proposal and other non-engineering (product/solution) aspects, who may also have been engineers previously, are usually in the minority. For example, in the case of a project to design a component for Royal Navy frigates, there was a ratio of 1:6, that is, one writer responsible for non-design aspects for every six design engineers contributing to the proposal. Twelve engineers worked on the technical part of the proposal, being directly concerned with the product, a piece of hardware, but only two worked on the rest of the solution, involving aspects like maintenance and other user-support.

The fact is, of course, that the 'solution' is the whole of the proposal, and includes a host of other considerations underpinning or impacting on the engineering design. At present, these 'other' sections are often compiled at the very final stages of the proposal writing process, by a small number of writers. Exhausted by the hectic weeks of working under pressure, the engineering design team seems to fade away leaving a (very) few of their non-design colleagues to finish the job. It seems that they feel their work is done. Sometimes just one technical author, possibly working with one or two others concerned with financial or after-sales aspects, is left to produce the finished document, which may be substantially larger than the technical section devoted to the engineering design.

Figure 9.8, which shows a breakdown of bid documents in terms of information content, presents findings that suggest there could be a case for distributing writing loads to better reflect the breakdown of the task. It is possible more writers need to be allocated to writing the other (non-technical) sections of proposals which, nevertheless, have a bearing on the overall solution. Proportions would seem to indicate a more desirable ratio within a writing team to be more of the order of 1:2 for the executive summary and 1:4 for the technical proposal, with the larger number favouring the design engineers in both cases. However, perceptions of the different types of engineers involved may account for the usual skewing of the writing teams in favour of the technical solution (see Chapter 9 for more on this). The figures also reveal that Integrated Logistics Support (ILS), or product maintenance and support, is an aspect of proposal writing that merits more attention and resourcing, especially with bids where this aspect is a major part of the customer's requirement.

This point is raised in Chapter 2, which discusses proposal writing guidelines issued by the customer. These may give an indication of task breakdown and where the emphasis of writing effort should be placed. It would seem that, if the customer stresses the importance of after-sales maintenance and through-life support, this should receive more time and attention in the proposal itself, with more support engineers being allocated to the task. However, proposals are never so simple. The technical solution may be a tricky one indeed, demanding the most intellectual and creative effort. The engineering work is where the greatest risk lies for a company. Underestimate the resources and engineering effort needed, or overestimate the system's performance, and the company is in trouble. In such cases, a company would lose even if it won: in other words, it would lose money in the long run, even if it won the bid. Each solution is unique and has to be judged on a case-by-case basis and, in the end, engineers take a pragmatic approach.

6.1.5   The Customer Requirement: a catalyst for proposal writing

Proposals are written in response to an identified need of the 'customer', which is a term of convenience used here to refer to the recipients of the documentation, that is, the target readers. In most cases, the customer is the person, or group of individuals, who has responsibility for reading and short-listing the proposals and, ultimately, for choosing the winning proposal. In the aerospace sector, the customer may be another similar organisation, for example, companies like Lockheed, Aerospatiale, Honeywell, BAE Systems, for whom a company may be bidding as potential subcontractor for a larger project, or the customer may be a government department within, for example, the Ministry of Defence in the United Kingdom or the Department of Defense in the United States.

A clear distinction exists between the terms 'customer' and 'user', since the two perform different functions with regard to proposals. The user may be the operator of the equipment, be he an able seaman or fighter pilot, but the customer is the person who vets the document, and, in a government department or commercial company, this would more likely be the head of a procurement team reporting back to an admiral and/or his aides, or the chief executive, respectively.

The original Requirement may take different forms: it could be an official document, like that described in Chapter 5, or it could be a hastily drafted email, an example of which is reproduced in Figure 6.1. This message arrived unexpectedly, asking for information about a product's performance and cost, the product in this case being a silicon gyro (SG). The sender needed the information in order to complete his team's own

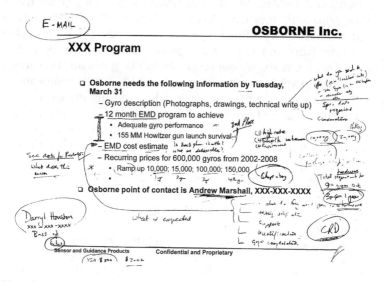

Figure 6.1   Unexpected email initiates proposal writing

larger proposal to a third party, bidding for a contract which would include, as a small part, an SG, or inertial measurement unit (IMU), as it is also referred to. This email message kick-started a cycle of events involving engineer members of the SG team, and, it so transpires, these events unfolded in a pattern closely corresponding to the product life-cycle stages described in Chapter 3.

The lead engineer immediately drafted his response, beginning with preliminary scribbles on the email printout itself. These notes later formed the basis of his email response, which led, a few weeks later, to the production of a formal proposal document.

At the time, the engineer saw this email request as part of a new trend, for his company was starting to receive many such urgent requests for information about this new IMU, giving his team little time to respond. He decided that an off-the-shelf generic proposal document needed to be designed, so that they could be better prepared in the event of future requests being made. His team's work on developing text for the IMU is described later in this chapter.

Research journal entry: Integrity and (possibly) misplaced honesty

I mentioned to Mark that I'd overheard them talking about risk a lot. Apparently they have decided to focus much more on price and compliance with this bid. Work on the best price, say that it is compliant, and sort out any problems arising later, is what they're saying. With previous proposals they've apparently been too honourable, explaining alternative options, if not fully compliant, and working out the best solutions, when actually the customer has a bottom line of price, delivery schedule, and compliance. I remember John Jeffries saying just that at one of our proposal writing sessions. Working on this bid, I think it is dawning on them that the customer is not being sincere. Either that, or the customer doesn't know what they're doing, so that the published criteria is window dressing, and that when they say 25% of their evaluation of the proposal is going on considering an alternative solution, it is simply not true.

All this second guessing makes the whole bid process really tricky, and they're getting into a sweat about it. Mark said little, since the proposal is hot, but I think they are going to be a lot more focused on telling the customer that he is getting what he's asking for and then arguing about it later. Anyway, what else can they do? It's not an off-the-shelf solution, but a tailored one where they seem to be bringing together various different packages and combining them into one product. They are concerned about being seen through the documentation as being honest and straightforward. I don't see how they would be compromising their principles by taking that approach. It would be a different matter if the product were off-the-shelf to any degree. But they do seem to be discomfited by the stance they've chosen to take.

Sincerity and honourableness – these are strong influences when they prepare their proposal documents. They smack a bit of old-fashioned attitudes which, I guess, may have been brought to the company by engineers who used to work for the armed forces [no swearing around me, chivalry, courtesy, etc.]. Even though not all of them served in the military, this high-minded principled approach seems to permeate the teams. I hope they don't lose because of it.

Research journal entry: The tension of waiting for news

Michael, Alex and Colin are finding the tension unbearable. They're still waiting for a decision about who has won the bid for the LAWD project. They went to London yesterday to answer yet more questions about through-life costs. The field has been narrowed to two, them and Ultram. They think they're in with a good chance, but won't know for another seven days because there has to be a cooling-off period. The stress is really getting to them. If they win, I'll get an invite to the party.

[A week later] Still no news about the LAWD proposal. They should have heard last Friday as it was the end of the cooling-off period, apparently. I rang from Birmingham to find out, but in vain. The prime contractor is moving very cautiously. Michael reckons they have to clear it with the customer, as any sensible prime contractor would, but that it is difficult to get in touch with people if everyone's on holiday.

Then Diane came round with a sweep for us to guess the day and time they would get the news! A pound each. I chose 10.00 on Thursday. Joseph Sennet has chosen Friday. People drifted up

---

Continued

during the day to place their guesses. Tension mounting. Dave Harris going around with a kind of rictus smile, telling me how he's finding the waiting hard to bear.

6.1.6 Economic and social impact of winning (or losing)

Since a proposal is written as part of the bid process, it has serious and formal connotations and, as such, is no different from any other proposal, be it a proposal made at a meeting, for example, or a marriage proposal; after due consideration, it may be accepted or rejected. There is happiness and celebration for the proposal team, if it is the former, and a sense of failure and dejection if the latter. As one bid leader put it:

> You put everything in to it. It's like going for a job interview: you've got to psyche yourself into the job, so that you actually visualise yourself doing it. It's a hell of a let down, then, if you don't get it. It's a serial process: when you submit a proposal you can see the project taking off and start to look beyond it to other spin-off projects. If you win, you have a party. If you lose, it's dreadful, because other plans fall by the wayside.

I have observed proposals being written by individuals and teams and, in the late stages, was working among engineers compiling a proposal for a gun system for the British Royal Navy. This was a particularly large proposal, in terms of the size of the actual document and the amount of time and effort expended on it. It was also a particularly important proposal, because, if successful, it would secure work for the engineers and those on the factory floor for a decade or more. If unsuccessful, the engineers knew that they would probably be disbanded, and that most of them could lose their jobs. The words quoted above were said by the leader of this bid team as they were about to make a team presentation to the customer, in the final stages of the bid process, after having worked on the proposal for over a year. In the event, they lost, and the dejection felt by the team was palpable. Soon after the news, they were steeling themselves for redeployment or redundancy; such are the human consequences of losing in the tendering process.

Research journal entry: Consequences of losing

Charles came by again to talk about the consequences of them not winning the bid. Basically, this value chain is EO [electro-optical], so if they lose the bid, ALL of this value chain would go. That means all of the engineers I've been working with in this section, including Joe, Martin Aspel, and Jerry. It also means that part of the factory floor would go too. They are fast becoming a silicon gyro site.

Research journal entry: After losing, to review or not to review?

If you lose the proposal, do you ever get together to review things to see where you went wrong? This is what I asked Harry. His answer was that they should, but they don't often do it. When I asked if it was because it was too painful for them, he didn't reply [engineers possibly find my language too emotional], but he smiled and repeated that they really should all get together to discuss it. Not just the engineers, but everyone involved in the bid. It strikes me that one of the problems with such a discussion is that it could lead to finger-pointing, fault finding and blaming people, which is not what the company is about these days. Espousing a team-working approach, they've long moved away from the blame culture of the 80s and early 90s. All the same he seemed to believe it was a good idea. I still think that with other proposals in the pipe-line, and the amount of time and effort yet to be invested, engineers don't have time to mope. They just lick their wounds and move on.

He gave another reason for not reviewing afterwards. Sometimes the selection process drags on and on. With CREST for example, two years on from the start of the tendering process, they are STILL waiting for a final outcome. So it's difficult to do a 'wash-up' when waiting for the results. They work to such long time-scales.

Postscript: The story of LAWD continues: the customer discovered that the company that won the bid could not actually deliver an effective system within budget, and so the proposal originally submitted by this company is being reconsidered. One engineer put it like this: 'the customer does not always tell you, at least not honestly, why he chose the other guy. Not all is fair in love and war.'

6.2 Persuasion

6.2.1 Engineers versus marketing colleagues

Engineers are unhappy about needing to persuade in proposals. They feel uncomfortable about it, and generally view themselves as being inept as persuasive writers. Winsor reports similar observations:

> As a profession, engineers frown on persuasiveness and find it suspect. (Winsor 1996: 12)
> The primacy and purity of data are an ethical as well as a functional concern. Thus engineers may believe they let the facts speak for themselves and abstain from any obvious persuasion because that is a useful fiction in the world of engineering. (ibid.: 99)

In the technical proposal, for example, they prefer to write in a style they believe to be objective and devoid of any emotive slant. They acknowledge that the main aim of a proposal is to persuade but do not want to be associated with the persuasive aspects of text or, rather, what they actually perceive a persuasive text to be. Engineers tend to associate persuasive language with salesmen, and overtly 'selling' language is somewhat offensive to them. One design engineer, for example, described an instance when he and the marketing member of the team did not see eye to eye about a technical description intended for a proposal. His recounting of the event demonstrates the cultural/attitudinal divide that exists even between those working in the same writing team. He was confident in his opinion that he should write about the solution using restrained factual language, about which no hint of a selling motive could be inferred. With a degree of hyperbole, he related their disagreement in the following words:

> The customer said he wanted it green, and so I wrote: 'It will be green.' But Michael [responsible for marketing] wanted it to say: 'You asked for green and you shall have it. You will have a beautiful shade of green. We love green at Matrix Industries. We have a whole range of greens for you to choose from', and I thought: 'I can't write that!'

6.2.2 Attempting to pin down the notion 'persuasion'

It would be a mistake to equate persuasion simply with 'sales talk', or with language that is subjective or hyperbolic. The prevailing attitude in the engineering community tends to reflect a dichotomy between objective/factual language and subjective/emotional language, and this rather black-and-white view is perpetuated in books aimed at an engineering readership. They emphasise the need for clarity and objectivity, but pay scant attention to subjective, loaded expressions or evaluative language (Bolinger 1980, Hunston and Thompson 2000, respectively).

Sometimes, there are obvious differences between language that is overtly persuasive and language that is not. The use of 'successfully' in the following two sentences is a case in point. In both, 'successfully' functions as an adjunct/adverbial, but in Example 1 it is used to emphasise and extol the virtues of the thermal imager, in a writing style that some engineers feel uncomfortable about, whereas Example 2 is a straightforward scientific statement of fact, explaining the algorithms used to engage the auto-track:

> Example 1: The Eagle 22 series of Thermal Imagers are designed and manufactured for use under the severest of operating conditions and have been successfully proven in ship borne and main battle tank applications.
> Example 2: When the processing circuits successfully differentiate an acquired target image from the background, then autotrack is engaged.

In an engineering context, however, it is useful to consider linguistic expression in terms of clines or continuums. This idea is not new: Houp and Pearsall discuss persuasive strategy in terms of a continuum, stating 'all information ranges along a continuum from complete objectivity to complete subjectivity' (1980: 141). They depict this diagrammatically (Figure 6.2), stressing the importance of the central 'Area of Reasonable Argument' in communication by stating:

> Many of our communications fall in this zone of the continuum, between pure objectivity and pure subjectivity. In this zone argument is permissible – in fact inevitable. And when we argue, we have at our command all other modes: exposition, description and narration. (ibid.: 142–143)

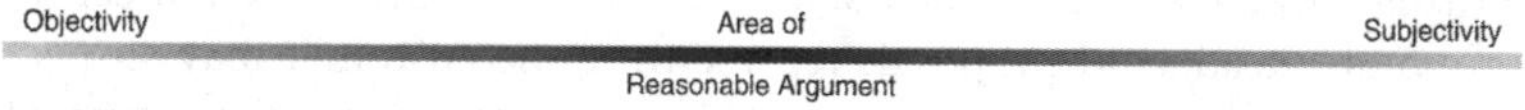

Objectivity   Area of   Subjectivity

Reasonable Argument

Figure 6.2 Persuasive strategy continuum (Houp and Pearsall 1980: 141)

However, categorical assertions about the existence of 'pure' objectivity or subjectivity would be absent from more recent views of discourse. These days, writers are seen to have a wider range of strategies and devices at their disposal, both linguistic and extra-linguistic, beyond the modes of exposition, description, and narration mentioned by Houp and Pearsall. Martin (1989) represents more recent (and to the engineers, contentious) views, sweeping aside any notion that factual writing is 'factual', putting forward the idea that factual writing requires creativity and imagination to be successful:

> There is a naive view in our culture that it is possible to distinguish form from content, and that factual writing deals with content and can be judged simply in terms of how truthful or close to the facts it is. . . . factual writing requires all the creativity and imagination we can muster if it is to succeed. . . . Exposition counts, even if it has nothing to do with truth. (1989: 49)

Possibly, one of the problems facing engineers is the confusion that results from the meanings associated with persuasion. As a term, 'persuasion' (or 'persuade') has rather wide colloquial applications, associated with the idea that it is an intentional act with illocutionary force (Austin 1975: 100). It is deemed successful if it results in the desired behaviour of the message recipient. Colloquially, 'persuasion' has potentially negative connotations, because of its association with social and behavioural manipulation, which is not benign. These associations have their roots in the views perpetuated by the likes of Packard (1957), who portrays a gullible society unsuspectingly persuaded into a pattern of consumer behaviour at the behest of clever advertising and unscrupulous corporations. Bolinger is one of the few linguists to explore aspects of persuasion, investigating persuasive acts which have negative connotations, for example evasion, instilling fear, persuading through deception, euphemism, and dysphemism. He discusses the lexical choices made to commit these acts, and is possibly the first to coin the term 'suasion', although he does not define it (Bolinger 1980: 110–111, 119–122).

It could be said that, since it may be considered as a kind of manipulation, presumably covert and possibly not in the message receiver's best interests, 'persuasion' is negatively coloured for engineers in certain writing (and reading) contexts. Let us take the case of engineering proposals. Readers of proposals representing the customer may suspect manipulative intent that is one-sided. They may think the proposal writers have adjusted the 'facts' in order to win in the bid process, and that these 'facts' have been skewed somewhat. Engineers generally regard non-technical English with suspicion (they refer to it as 'natural' English), because it allows, in their view, vagueness, 'truth-bending', and inaccuracies. It is possible, therefore, that by acknowledging that proposals persuade, engineers make the task of writing more difficult because of the problematic association with potential dishonesty and malign intent. It is indeed a conundrum that faces the engineer, since he has to convince the reader of the proposal's benefits to the customer, while at the same time both reader and writer are aware that the proposing company also stands to benefit. Not only that; from the customer's viewpoint, the proposing company may benefit to an even greater extent.

6.2.3 Engineers' ambivalence towards overtly persuasive language

So far as engineers' views on persuasion are concerned, then, the picture is not clear. Numerous discussions with them over the years have shown, though, that proposal writing has demanded that they have little choice. At times, engineers need to produce persuasive text, although they work hard at not being too obvious about it. They are not alone. There are others in the scientific field with the same burden of having to write proposals to ensure the continuation of their professional practice and livelihoods. Myers mentions some of the constraints placed upon biologist researchers in their attempts to bid for funding. He describes a similar situation that exists for the biologists:

> There is a paradox in the rhetorical strategy of the proposal, because the proposal format, with its standard questions about background and goals and budget, and the style, with its passives and impersonality, do not allow for most types of rhetorical appeals; one must persuade without seeming to persuade. (Myers 1990: 42)

It seems the engineers are correct in believing that an informal writing style or an overt 'sales pitch' is generally disapproved of in the engineering field. Readers of proposals at the MoD, for instance, who are themselves engineers, say that they also react negatively to such a style of writing. The engineers are faced with a problem, then: how can they write persuasively in a style acceptable to their peers? This is indeed a bind they find themselves in, since the readers of their texts are usually other engineers. All have been taught by their science and university teachers to write in a formal and impersonal style. My work with engineers in commercial organisations and academic institutions has shown this attitude to be ingrained in the engineering community. This observation is confirmed by Kirkman, who discusses the 'traditional' writing style produced by engineers, describing much of it as 'heavily unreadable':

> ... when I suggest that passive, impersonal, turgid expression is a millstone that the technical content need not carry, I am told that papers written in any other style would be unacceptable: 'It would be thrown straight back'; 'My boss wouldn't have it'; 'You must make your work sound impressive' .... Always there is anxiety that other engineers and scientists would not accept a departure from 'traditional style'. (Kirkman 1992: 2)

However, in recent years, in small pockets of the engineering community, attitudes amongst engineers are softening, as they realise that their livelihoods rely on it, especially where engineering proposals are concerned. Nevertheless, there remains disagreement about exactly how proposals should be persuasive, with ambivalent attitudes still prevailing towards the notion. Some engineers make uneasy attempts to be persuasive with gauche results. The following enthusiastic self-endorsement, for example, elicited only raised eyebrows and a sceptical response from the customer; in this case, a procurement team at the MoD. Unfortunately, the use of 'excellent' (twice in a short paragraph) together with the claim about a successful working relationship evoked in the readers the opposite effect to that intended by the writers:

> SPACETRONICS believes this program is an excellent match for its advanced IMU product and business plans in terms of performance, price and quantity. Equally, we believe that the excellent working relationship built up between our two companies, even before the merger, during the initial bid phase has demonstrated that working together can be successful.

Since engineers are not traditionally associated with such rhetoric, little help is available in the literature, either within engineering or applied linguistics. English for Science and Technology (EST) writers (Hicks 1961, Pauley 1973, Fear 1977, Kirkman 1992, and others) tend to assume that, in the main, engineers write factual, information-conveying, non-argumentative texts. Understandably, they deal with text-types more usually associated with engineers, for example engineering instructions, specifications, and reports. As a result, their central concern is to encourage accurate, factual writing, perpetuating the notion that this is what engineers usually write, and that they need to concentrate on specific aspects, like writing clearly, concisely, and objectively. It is understandable then that engineers are uncomfortable about any writing that differs from this, and are usually exercised over how to include information that will persuade prospective customers.

There is a dearth of information available about the persuasive kind of language engineers use to do this, though. Chapter 8 discusses advice they are given about this, which advocates identifying particular persuasive features. Engineers are advised, for example, to persuade by describing different selling points or themes, in the case of proposals. However, the language used to express the themes, including such phrases as 'ensure low risk', 'engineering excellence', and 'committed to quality', are regarded as unsubstantiated 'sales talk' and stylistically undesirable by engineer proposal writers and readers alike. Other kinds of advice, for example, the writing guidelines for proposals issued by organisations like the European Space Agency, the American Military, and in-company guides, tend to concentrate on formatting rather than language expression.

Research journal entry: To blind with science or not? That is the question

Sometimes, deciding what to include or not gets too complicated. My chat with Brian, Steve and Jeremy shows just how double guessing about craftiness can take you places you don't really want to go when you're writing. We were talking about an uncompetitive, solicited proposal (so the customer actually wants it and is clearly keen to see it), but even so they are still stressed about what to put in and how to put it. What I gathered from this conversation is that engineers really want to impress the reader but don't want to be seen to be crafty, and so they get involved in compositional contortions and complex decision-making about what to include. No wonder there is so much angst about proposals.

All I did was mention the bit containing the stats and science/maths-'speak'. I asked them about this curious section of statistics that seems out of kilter with the rest of the document. It is written in a completely different style and makes for a different reading experience [for a non-engineer like me]. They said it was there to impress the customer with the high level of scientific

Continued

thinking and research facilities implied by the calculations, and access to an organisation like Sowerby. Brian's actual words were that he wanted the customer's reaction to be: 'My God, it's come from Sowerby. That's like Oxford or Cambridge in the university world. It's got to be good.' He then said (somewhat sheepishly) that a few of the readers, even though engineers themselves, may not understand the calculations! So what's the point of putting them in then? Apparently they are trying to prove mathematically that the thermal imager has a 20% improvement in performance, and of course engineers are rarely persuaded by anything that isn't 'factual'. So they need to include this kind of information. Without any prompting from me, he said he reckoned the mathematical explanation should be in an annexe, and a summary of the improved performance in the main proposal. I agree, absolutely.

But then Steve mentioned that, if the reader doesn't understand the calculations, the text may have the opposite effect of the one intended by eliciting an unenthusiastic response. He said it could make the reader silent, because he doesn't fully understand the calculations. They could even make him think they'd been included as a smokescreen or camouflage for vague claims about the product [or 'vaguity' to use Steve's word]. And as he points out, these are all theoretical claims, because the proposed thermal imager is unproven in the field as yet. So there are times when engineers need to be vague and, when they are, they use arcane technical information [and language] to hide behind. They know it, and their engineer readers know it. The danger for this proposal is the reader may suspect this as being the case when it actually isn't!

6.3 Engineers' attempts at restrained persuasion

6.3.1 Restrained persuasion: Example 1

All the same, there are engineers who will admit to composing text with the aim of persuading the reader. The engineer in the following example is a support engineer, conscious of the need to build a relationship of trust with a potential customer, who spent around two hours trying to compose a few 'right-sounding' sentences. Making slow progress, he approached colleagues sitting near him for their renditions, but when their offerings were not what he wanted, he asked me for help. At first, his task seemed a simple one: he was composing a brief written response to a few of the many numbered requirements in the Customer Requirement, in particular, two which read as follows:

5.1.5 Maintenance at Test Environment shall be supported by the Contractor.

5.1.6 The implementation of any upgrades will be undertaken on an opportunity basis to minimise the impact on the operational programme of the warship and Test Environments.
[ILS requirement]

He was trying to write a sentence as part of a longer list of goals, or stated intentions in a proposal, wanting to convey, above all, sincerity in expressing a desire to please the customer. He wanted to impress the reader with the company's commitment to providing a quick and efficient service, especially at times of risk during the testing phase of the product. He wanted to convey the fact that the company would not only provide upgrades when they became necessary, but would be so efficient and aware of the customer's needs, that problems would be anticipated even before they occurred, and, furthermore, that engineers would suggest improvements and upgrades before the customer realised they were necessary. How, he wondered, could he 'reach out' and impress the customer with this complex message of sincere intentions, while at the same time, to use his words, make him 'have confidence in us, and make him feel good'?

It transpired that he was looking for 'right' words, but was finding this to be a problem because, in his view, he had a 'vocabulary deficit'. However, he knew the nature of the (emotional) response he wished to evoke in the reader, and also had a vague idea about the language he was prepared to use in order to achieve it. He had a clear idea of a preferred style, wanting to convey the company's professionalism and commitment to a rapid response without it sounding too 'slick' or glib. In spite of the emotive spoken language he used in the conversation to talk about what he wanted, he was not prepared to sacrifice received stylistic norms in his writing. For example, of the ideas bandied about, he rejected 'rapid response' as being too clichéd, and 'speedy response' as 'sounding like a plumber's ad'. I agreed to compose a few 'straw man' sentences for him to consider, and emailed them to him. Here is part of his final response to those numbered require-ments, of which the only words of mine that he used are 'proactive support':

1. To identify and develop a maintenance concept which limits main-tenance at sea, thus ensuring the least effect on the operational avail-ability of the warship.
2. To provide a proactive support policy for the Neptune 22 at the test environments to prevent delays to the programme.
3. To implement any upgrades to the Neptune 22, in an effective manner, as dictated by the operational programme of the warship and test environments.
   (Taken from 'ILS Goals')

Because they are listed in this way as parallel structures, the sentence adjuncts (providing a reason or purpose for each proposed item) are in final position, the adjuncts in these sentences being 'thus ensuring the least effect on...', 'to prevent delays to...', and 'as dictated to by the operational programme of...'. I have noticed engineers' avoidance of the use of adjuncts, when describing the product (or solution), and the almost total absence of them in sentence-initial position. The significance of this observation could be that engineers avoid using adjuncts, since they instinctively sense that adjuncts convey, to their way of thinking, less factual information, but larger samples need to be examined to see if this is a general tendency and worth further investigation.

The above extract epitomises the kind of writing style considered by engineers to be acceptable persuasive language. Each is listed as a separate item, and focuses on particular aspects of the maintenance provision considered by the support engineers to be of most concern to the customer. Each relates to a specific item of information, paraphrased below to show the meaning of each:

1. Minimise inconvenience to the customer by reducing the mainten-ance that will have to be done at sea.
2. Anticipate problems in order to pre-empt them.
3. Conform with the ships' working schedule, so as to avoid interrupting planned operations.

In the engineer's version, negatively coloured or minimising words are employed to convey positive attributes modestly, for example 'limits maintenance' and 'the least effect'. The most positively loaded words in the piece are 'maintenance concept', 'operational availability of the warship', 'proactive support policy', 'upgrades', and 'effective manner'. Any more effusive expressions would not pass muster.

6.3.2   Restrained persuasion: Example 2

The following text segment is part of an opening paragraph for an executive summary. Executive summaries accompany technical proposals, summarising the main selling points of the product being proposed (explained in more detail in Chapter 9).

> *Development Experience in Naval Electro Optical Tracking Systems*
> The Neptune 1 Electro-optical gun fire control system was developed in the 1970s. Sea Neptune 1 systems were fitted to the Royal Navy TURRET class patrol vessels and exported to customers in the Middle and Far East. Using this experience in naval Electro-optical tracking, OSBORNE INDUSTRIES developed the high performance All Purpose Electro-Optical Director (APOD), which was selected by the Royal Navy in 1980. The resulting Neptune 22 system (designated DASS in the Royal Navy) provides Electro-optical surveillance, tracking and gun fire control.

The heading 'Development Experience in Naval Electro Optical Tracking Systems' would be considered unremarkable by most readers, who would see it as fairly representing content to follow, which is, after all, a list of specific examples of product development experience, providing a historical outline. The ostensible purpose, then, is to provide historical information about the product's development. This may cause some puzzlement, however. Bearing in mind that the aim of the executive summary is to persuade the reader about the benefits of the proposed product, the reason for including this historical background is not apparent. What could be the reason for including this historical segment, and what particular persuasive point were the writers trying to make? After all, a potted history of a product is not inherently persuasive. Clearly, inferencing skills need to be exercised here to tease out of the text some less obvious purpose.

A positive feature, in this case 'proven performance', can be inferred from the extract, which is, essentially, an attempt to convince readers about the effectiveness of the product. By inference, the fact that the product is already in use by the Royal Navy and other navies could be considered to be proof of performance. An interview with an engineer confirmed this impression. He explained that 'proven performance' is a term often used to refer to the numerous tests the product has been submitted to and passed. This is a positive feature worth advertising in executive summaries. This is understandable since the product in question, the All Purpose Electro-Optical Director (APOD), must have been used by the Royal Navy for decades, since the 1970s. As part of design procedures, the APOD would have been submitted to a range of tests over the years as a matter of course, and its design modified as a result. In a sense, the words 'Proven performance of' would seem more applicable than 'Development experience in' in the heading, but this would doubtless prove to be unacceptably boastful or too direct for some engineers.

To illustrate their sensitivities over this sort of language, when asked about the piece, another engineer agreed it concerned proven performance, but, after further mulling, introduced other terms. He suggested, for example, 'battle hardened' to describe equipment which has proved effective in a battle situation, adding it would have either proved itself or had its design refined using assessments of its performance in an authentic battle situation. However, such a label could have only limited use, since not all the products designed by this company are produced for use in battle. Probing further and searching for other headings to help categorise the extract, I suggested 'Broad customer base'. His response was, 'That might imply satisfied customers, but they aren't always satisfied.' Ultimately, accuracy of information is paramount and, with these engineers at least, there can be no bending of the truth.

The point of these discussions was to help identify text elements in order to undertake a systematic analysis of executive summaries and technical proposals. It emerged that engineers readily accepted the persuasive intent underlying these documents, but they were not always clear exactly what they needed to be persuasive about. The results of the investigation into the 'aboutness' (Marder 1960: 61) of these texts are discussed later (viz. Sections 8.4 and 8.5).

6.3.3   Restrained persuasion: Example 3 – composing text for a generic proposal

This example is based on the recent work of a small team of engineers working for a company that has developed a new product, which we shall refer to as the SG (see p. 131). At the start of this century, the SG was a new product, completely unlike the mechanical spinning gyroscope that was its predecessor. As it was improved, its usefulness and versatility became apparent to SG engineers, who became increasingly aware of the potential for growing the business in the area of SGs. They sensed there was an urgent need for suitable publicity material to help expand the market for the SG and realised suitable descriptive text should be available to send to potential customers, or as responses to RFIs. In fact, both the engineers and sales people at the company believed there was potentially a huge market for this new generation of 'gyroscope', if only people knew more about it.

Engineers thought there was a lack of understanding (of a scientific and mathematical kind) about SGs on the part of the general public, who did not know what they actually were or how they worked, and general ignorance about SG applications in a variety of contexts and machines. They also realised that potential customers in other companies, who may be interested in the SG for incorporating into their own products, also probably lacked the necessary scientific/technical knowledge to fully appreciate the ingeniousness of the design or its (potentially) multifarious applications.

It was dissatisfaction with earlier attempts to produce proposals with a clear 'selling' message that motivated the team leader for the SG, James, to set up a task force to develop a generic proposal. He was receiving numerous RFIs and needed a document that could be sent in quick response, without taking up too much time of the engineers who were busy with SG design developments. Marketing colleagues had produced a text, entitled 'Micro-machined silicon puts a new spin on gyroscopes' (referred to from now on as the 'New spin' text), to be used as the product description in publicity material and proposals, with the aim of making the SG more appealing to a wider audience. James called a meeting of the SG engineering team to discuss the text's suitability for inclusion in their generic proposal. The marketing and sales team were unaware of, and would have been surprised at, the reaction their text had engendered amongst the engineers.

At the very start of the meeting, James made clear his disapproval of the 'New spin' text, saying he was determined to replace it, because, to quote his words, 'It's not up to the job, and it won't do.' He and his engineers expressed some embarrassment that it had already been included in a few proposals, describing it as being 'too snazzy' and 'brash'. They expressed dislike of what they considered to be the overt sales pitch of the piece. The first page of the offending text (which is four pages in length) is shown in Figure 6.3, and a few extracts from it are included below to illustrate features the engineers found distasteful. First, they objected to the title, dismissing it as similar to a tabloid headline. They described the style as being too 'chatty', but could not be more specific about particular 'chatty' features of the text. After some discussion, it appeared they objected to certain lexical choices, the structure of the sentences, and punctuation. As an example of a structural

MICROMACHINED SILICON PUTS A NEW SPIN ON GYROSCOPES

Helix Industries
Systems & Equipment

HISE, the leading UK vendor of gyroscopes has announced an industry first - it has developed the world's first commercially available micromachined ring gyroscope.

Over the past few years, developers of inertial systems have placed some important demands on manufacturers of gyroscopes. Not the least was that gyroscopes should be smaller, more reliable and less expensive. The result was that gyroscope designers moved away from a purely mechanical approach towards solid state devices. Today, these solid state designs have been in successful production for some time and have been used very effectively for platform stabilisation, remotely piloted vehicles, pointing, navigation and control systems.

But the market never stands still, and today, new requirements are driving the technology for the next generation of gyroscopes. This time around, it is not the military market that is asking for a low cost gyroscope system, but the commercial marketplace. Developers of vehicle dynamic control systems, navigation systems and active suspension systems all need gyroscopes which are more than an order of magnitude less expensive than their present day counterparts.

To address these needs, gyroscope designers have no alternative but to embrace new manufacturing technologies. In effect, this means that they must turn to micromachined silicon components for the mechanics of the gyroscope. They must then closely integrate the micromachined result with control electronics built from a single Application Specific Integrated Circuit (ASIC) in order to meet the required price/performance criteria.

In the development of these micromachined gyroscopes, techniques and technologies previously employed in the design of earlier systems can be

effectively employed - and then cost reduced.

Such was the case with the development of the new silicon micromachined vibrating structure gyro from Helix Industries Systems and Equipment (HISE). Now in production, this new gyroscope will be delivered at a price an astonishing order of magnitude less than was achievable from older generation designs.

With the development of its earlier VSG2000 gyroscope, HISE designers developed an innovative patented mechanical ring structure to take the place of previous more expensive, mechanical rotating structures with gimbals and bearings. It was certainly cheaper. This new solid state gyroscope worked by making use of the Coriolis force - a force that is observed to act to a moving element in a rotating body.

In operation, the HISE mechanical ring gyroscope was made to oscillate by the application of alternating force to the ring. When this oscillating body was placed in a rotating reference frame, the Coriolis force came into play, producing a secondary oscillation orthogonal to the primary oscillating motion. (Figure 1).

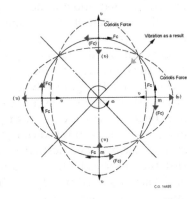

Figure 6.3 The 'New spin' text (first page)

objection, this sentence was anathema to them, because they would never include sentences beginning with 'but' in a proposal:

1. *But the market never stands still, and today, new requirements are driving the technology for the next generation of the gyroscopes.*

They also would have preferred some other verb to 'driving'. Another aspect of the kind of sentence feature they disliked was, in particular, the frequent use of sentence initial adverbials/adjuncts, like 'Now in production' in 3, and the use of dashes (next example), which are used three times on the first page. In this respect, they would concur with Quirk *et al.*, who advise cautious use of the dash because of its dramatic (and one would infer, distracting) effect and the informal impression it conveys (1985: 1629):

2. *In the development of these micromachined gyroscopes, techniques and technologies previously employed in the design of earlier systems can be effectively employed, and then cost reduced.*

However, their strongest objection was to the phrase 'an astonishing order of magnitude', particularly the use of 'astonishing', in this sentence:

3. *Now in production, this new gyroscope will be delivered at a price an aston-ishing order of magnitude less than was achievable from older generation designs.*

They naturally objected to the hyberbolic 'miracle' in this sentence:

4. *The electronics in the single ASIC controller is also a miracle of modern levels of Very Large Scale Integration (VLSI).*

They also considered their portrayal in this sentence as overly dramatic:

5. *To address these needs, gyroscope designers have no alternative but to embrace new manufacturing technologies.*

Judging by their comments, engineers tend to avoid using adjectival intensifiers or lexical adverbs of the kind used in the text, for example 'Silicon has some very useful material properties for a sensor', 'The gyroscope has also been carefully designed...'. The 'New spin' text is actually a creditable attempt by non-engineers to portray the innovative and ingenious features of the product in an interesting way. With some work on stylistic aspects, it might have been converted into a more acceptable version for use by these engineers in their proposals. However, the original motivation for writing the piece, with the notion of a one-size-fits-all text intended for disparate audiences, is basically problematic. Clearly, if a text concerns a description of the product and reads like a press release, it is unsuitable for inclusion in a formal proposal. In the event, such was engineers' distaste for the writing style of the piece that they rejected it out of hand, and decided to write their own version from scratch.

This example serves to illustrate that engineers believe they more effectively persuade by describing the product using a kind of scientific writing more formal than that used in the 'New Spin' text. They prefer technical writing that includes the language of definitions and other scientific (or technical) descriptions, incorporating explanations with equations and formulae. Here are just a few typical examples:

Definition:

Gyroscopes are instruments which are used to measure angular motion. The Vibrating Structure Gyroscopes described in this paper are simply devices which provide a voltage proportional to the rate of turn applied to the gyroscope's sensitive axis.

Technical description:

The cylinder is manufactured as a single part, it is closed at one end which has a stem for mounting purposes. In order to provide the means of driving and detecting the vibration, electrodes are printed on the cylinder. These are equally spaced around the circumference of the cylinder.

Expression of formulae:

Mathematically the coriolis force (Fc) is equal to twice the mass (m) times the vector cross product of the angular rotation frequency (ω) and the linear velocity of the mass (v). That is, $Fc = 2m(\omega \times v)$

On occasion, their descriptions take account of the fact that readers may not have much technical knowledge. They are fond of using analogy to explain scientific phenomena, tending to use the car or parts of a car to explain a point. However, a garden hose pipe and a playground roundabout are the preferred analogies to explain the coriolis force in the next example:

Analogy:

The simplest example of this coriolis force is to imagine watering your garden with a hose pipe: as you turn the water expelled from the pipe appears to move in a curve. In the fixed earth bound frame of reference this is due to the positional lag due to the rotation.... If you don't believe this is a real force, the next time you visit a children's playground try kicking the centre of the roundabout while it is spinning, and you are on the ride, and you will feel the force pushing the foot away.

This is a rare example of more informal, personal language, with the use of the personal pronoun 'you' and reference to leisure-related aspects, such as 'your garden' and 'children's playground', which engineers usually avoid using in technical writing. It shows an attempt to 'reach out' to the layperson to explain the mysteries of the coriolis effect in everyday life. The fact is engineers will happily compose text for sales and marketing purposes, so long as it is expressed in language they believe would be acceptable to the engineers in other companies who would be reading it. As Chapters 1 and 4 explain, such language must be restrained, formal, and give the appearance of being objective. The following two extracts show how engineers achieve this.

Extract 1: It is felt that in the future most conventional motor driven, spinning-wheel gyroscopes will have been supplanted by solid state devices. This is due to their simplified construction, improved reliability and shock handling capability.

Extract 2: Short Run-Up Time – The sensor has no conventional spinning wheel and, as a result, achieves ready state within a very short period. This period is governed only by the capture time of the phase locked loop used to excite the resonant structure, and is in order of 300ms. This time can be adjusted according to requirements.

They use long complex noun phrases, which in these extracts are mostly heavily pre-modified, for example 'most conventional motor driven, spinning-wheel gyroscopes', 'shock handling capability', 'the capture time of the phase locked loop used to excite the resonant structure', among others. The passive verb phrase, as ever, is one of the features of formal expression, for example 'is felt', 'will have been supplanted', 'is governed', 'used to excite', and 'can be adjusted', as are other phrases associated with formal language, for example 'this is due to', 'it is felt that', and 'according to'.

6.4   Concluding observations

The contrast between the drama of the bid process (for that is what it is) and the cool formality of the objective language engineers wish to use is striking. Everyone in the bid team, including the technical authors, sales staff, commercial and legal experts, and people in the print room are caught up in the rush of events, experiencing along with the engineers the pitfalls and excitement that are part and parcel of producing a proposal document. Bid preparation consumes huge amounts of energy, expertise, and time, to the exclusion of other work activity. All is focused on getting the proposal documents produced and delivered on time. These are big texts, and, as is the way with such texts, they cause stress and bring drama to those who work on them. Engineers may be reluctant to use the expression, but others have no hesitation in describing it as being 'mayhem' at times.

More important, then, engineers would claim, to steer a steady path by upholding engineering values in the processes and procedures relating to textual matters. This inclination is deeply ingrained and it really matters to them that they should conform with the linguistic conventions of their discourse community. They are particular about how they describe their products, not only because they wish to be as accurate as possible, but because they are displaying their credentials, their knowledge, and expertise through this text. The persuasive message is complex because the text is in fact 'selling' the engineers as well as the product. In other words, they are attempting to 'sell' their products by displaying their own knowledge and expertise via text. So the image the engineers wish to project of themselves to those they are targeting (engineers working for other companies) is portrayed, and substantiated, by the style of writing and informational substance of the description.

Thus, proposal documents provide an opportunity for engineers to convince readers of their expertise and high standards of professional practice. They are opportunities for 'giving face' to engineers, who wish to convey these qualities about themselves (and the company). They usually feel strongly about this, but find it hard to describe. Later chapters explain how they try to achieve this through presentational factors, like choice of cover, pictures, colour, and so on. More importantly, engineers attempt to impress the customer with particular types of information, referred to as proposal components (PCs) in this book. These are also identified and described in later chapters.

7
The Presentation of Engineering Proposals

7.1 Introduction: textual cosmetics

Technical authors occasionally lament the time and effort engineers put into word processing their contributions. They spend inordinate amounts of time, as the authors describe it, not only on their written compositions, but also on using special word processing features to 'prettify' them. It seems they cannot resist using special fonts, unusual table styles, and other layouts to present their work. Inevitably, when the various pieces are collected from different members of the design team and collated by the authors, the texts are stripped down, so to speak, reformatted, and reorganised to conform with the overall style and structure of the whole proposal document. Along the way, the authors may rephrase the odd sentence or rewrite whole sections, which usually receives a good reception from the engineer writers, who like to see their text improved in this way. On occasion, however, the changes they make are greeted with protests from the offended engineer writers, who clearly dislike seeing their work changed. Certainly, some engineers show a proprietary attitude towards their writing, at times 'fiddling' with it to return it to its original form, even when reviewing later versions of it that have been changed by the authors. This textual 'sparring' reveals a tension over text ownership and control, that leads to rising passions and lively exchanges (rarely acrimonious) about proposal text. It would seem there is a need for better understanding and communication between technical authors and engineers about work roles. However, this also demonstrates the importance engineers and authors attach to the presentation of proposal text and their commitment to presenting it in the best possible light.

7.2   Physical features: size, formats, and outline structure

7.2.1   The textual 'extent' of proposals

Perceptions of size are relative. Much depends on the size of the company and views about the potential remuneration of the contract being bid for: a proposal of around 4000 words in length would be considered small in some companies, and be bound as a single document, as depicted in Figure 7.3. On the other hand, another proposal may comprise several volumes amounting to hundreds of thousands, or even millions, of words. Such proposals usually comprise compilations of sections written by different authors, the composition process being not unlike the construction of a jigsaw puzzle, where pieces are fitted together to form a whole entity. The 'pieces', so far as proposals are concerned, are text segments (or components) distinguished by their information topic. Chapters 8 and 9 discuss the notion of information topic and suggest a taxonomy of topic-based components, called PCs, that make up the structure of technical proposals.

Table 7.1 shows the breakdown of a proposal comprising over 156 000 words, c.58 000 of which form the technical proposal section, the part written by engineers. Engineers would describe such a proposal as

Table 7.1   The textual 'extent' of a fairly large proposal

Description			No. of words in section	No. of pages in section
Volume 1 – Commercial Proposal				
EP3049 Volume 1			5,496	57
		Total	**5,496**	**57**
Executive summary	(included electronically in only Volumes 3 and 4)		1,661	8
		Total	**1,661**	**8**
Volume 2 – Project management proposal				
EP3049 Volume 2	Section 1		2,728	17
	Section 2		3,791	16
	Section 3		2,395	16
	Appendix A		37,976	144
	Appendix B		12,207	50
	Appendix C		664	5
		Total	**59,761**	**248**

Volume 3 – Technical proposal

EP3049 Volume 3	Part 1			8,892	41
	Part 2	Section 1		440	4
		Section 2		42	3
		Section 3		5,293	31
		Section 4		9,209	35
		Section 5		4,653	36
		Section 6		9,505	39
		Section 7		5,117	25
		Section 8		2,083	11
		Section 9		234	3
		Appendix A		3,685	21
		Appendix B		224	3
		Appendix C		8,368	44
		Appendix D		870	9
			Total	**58,615**	**305**

Volume 4 – Tender support technical data

EP3049 Volume 4	Section 1		53	3
	Section 2		240	3
	Section 3		987	5
	Section 4		173	3
	Section 5		303	4
	Section 6		5,996	27
	Section 7		5,773	18
	Installation specification		17,628	101
		Total	**31,153**	**164**
OVERALL TOTALS (excl. Commercial Response)			**156,686 words**	**782 pages**

consisting of a 'set' of documents, only one set of which would be their primary concern, that is, the technical proposal. Figure 7.1 is an extract from another proposal, considered large by the company producing it. In view of its size and complexity, the technical author responsible for compiling it included this 'route map' for the benefit of the readers. On the left is a list of what the customer asked for, and on the right, where these could be found in the proposal. Having in mind both those who would be in overall charge of the reading process, and those who would read particular sections of it, he included a copy of the 'route map' in each of the four volumes. Similarly, as can be seen in Figure 7.1, each volume contains a copy of the executive summary, to provide the specialist readers of each volume with an overview of the main benefits of the proposal.

Since engineers may contribute only a section, the sheer size of such proposals may escape their attention. In the case of the proposal depicted

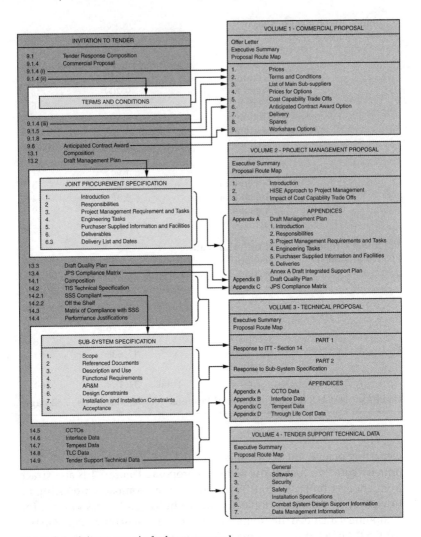

Figure 7.1   A 'route map' of a large proposal

in Figure 7.1, for example, only the technical authors and bid leader saw the final document when it had been printed and bound. Two engineers, who had contributed significantly to the sections on the design, mentioned that they had not seen the final result and did not know what it looked like. Table 7.1 and Figure 7.2 provide a visual impression of the 'textual' extent of a large proposal. Table 7.1 lists all the sections

Figure 7.2 Paula from Tech Pubs carrying a proposal

in the four volumes of the proposal, plus the executive summary, and the number of pages and words each contains. As can be seen, the proposal comprises a total of 156,686 words and 782 pages. Following the table, the photograph Figure 7.2 provides a pictorial impression of the size (and weight) of the submission: it shows a photograph of a member of the technical publications team, referred to as 'tech pubs' in the company, carrying the four volumes that were submitted to the customer.

Research journal entry: Few see the end result

A bit of excitement today. And all because Tony D-E asked a deceptively simple question about Proposal X. He wondered if I could be more specific about the physical size of it. Paula offered to do the

---

Continued

donkey work of trawling through each individual folder to count pages and words. She caused ripples of interest when she went round the team asking for hard copies of the proposal and the engineers willingly cooperated, ferreting about in locked cabinets and cupboards for bits of it. She became weighed down with all the volumes, so I decided to take a picture of her carrying them. Tony was fascinated to see how large the proposal was, which made me realise he didn't know what the whole thing looked like, even though he'd had a hand in writing it. That's a thought: in the rush to get the thing written, the only people to see the end product before it went to the customer was probably the technical author and the print room!

Knowledge about what happens when a proposal is delivered to the customer influences the way it is structured. We know the 'customer' is more than likely to be a collection of teams of specialist readers, each team responsible for one of the four sections. Therefore, if it is thought (the bid team can only surmise about this) that the proposal will be automatically split up into sections and distributed to different readers with different reading responsibilities, it makes sense to structure the proposal to facilitate this process.

Typically, a proposal can been seen to comprise four key sections, or volumes, depending on the size of the potential business, as shown in Table 7.2. A variety of sections (and section headings) make up each volume and, depending on the contract or type of solution being

Table 7.2　Typical breakdown of a proposal

Section or volume	Information content
Executive summary (concerns the engineer)	Synthesis of the main 'selling' points of the bid
Technical section (concerns the engineer)	Company's technical response to the customer's Requirement, including technical description of the product, compliance matrices, maintenance of the product, and measures for ensuring its successful performance when in use

| **Management section**
(not of direct concern
to the engineer;
involves other writers) | Administration of the programme, production plans,
the company, and personnel |
| **Commercial section**
(not of direct concern
to the engineer;
involves other writers) | Financial and legal aspects; commercial terms and
conditions, including prices |

proposed, different names may be given to the separate volumes of the proposal. Other sections may be included, as shown below, but the core elements of any proposal are shown in Table 7.2.

7.3 Significant parts of the proposal

7.3.1 An overview of each section

Figure 7.3 shows distinct sections of a typical proposal pictorially. The sections with bold labels are examined in more detail in this chapter. The title page, glossary, and proprietary statement may not appear to

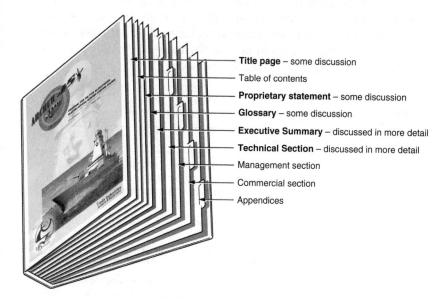

Figure 7.3 Pictorial representation of proposal sections

be obvious choices for a closer examination, but have been included in this description for three main reasons:

1. They have distinctive visual features, performing particular functions relating to the whole of the proposal. A closer examination of them will provide useful contextual information for understanding proposals.
2. They are integral parts of every proposal, and are always placed at the front, together with the table of contents. The 'fronting' of these parts has early utility and impact on the reader.
3. They provide important information about the proposal relating to the way it is regarded and the manner in which it is read. They show the engineers' communicative intent, and help to prime readers by imbuing them with the desired attitude. Put simply, engineers and technical authors use these sections to influence how readers approach the reading of the proposal, the way they read it, and how they should regard the ownership (copyright status) of the information it contains.

7.3.2 Generic outline structure of proposals

A tabular breakdown of all proposal sections and discourse functions is provided in Table 7.3 although not all parts of the proposal are given detailed examination in this study, as explained above. This breakdown arose out of analysis done to provide the basis for devising a generic proposal at the request of a team of engineers.

Table 7.3 Generic outline structure of engineering proposals

Main sections	Discourse functions
Front cover of folder and/or title page	*Names/Refers* [Referential and/or nominative function] – names the proposal, the company, and the key personnel who are responsible and/or accountable for the bid. Aims to: a. help organise the reading of it, administration, and storage b. impress the reader by its appearance and visuals c. provide reader with an idea of the nature of the product.
Copyright page/ Proprietary statement	*Asserts (and establishes) legal copyright* of contents of the proposal based on the company's proprietary statement

Executive summary [sometimes in letter form]	*Persuades* – ultimately, sets out to persuade the reader to place the proposal on the shortlist, by providing one or more of the following: • overview of the main selling points (called 'themes') of the proposal • description of the product • information about the company, intended to impress.
Table of contents	*Facilitates reading process* – Reading team use this to: 1. gain an overview of the document structure, in order to 2. split the document up to be read by different members of the reading team, and the more conventional use, which is to 3. help the reader navigate and read the document.
Glossary	1. *Provides information about terms and acronyms* so that readers can read the text and (numerous acronyms) with better understanding. 2. Provides writer with the means of avoiding repeating names or phrases to produce more readable prose [although this aim is not always fulfilled].
Technical proposal	*Impresses the reader* (or attempts to) by explaining the product from different engineering perspectives, for example, describing: • key aspects of the design • degree of compliance • physical and functional characteristics • comparisons and contrasts with other like products • manufacturing aspects • testing and research • the company's expertise/ track record.
Commercial proposal/aspects of the proposal	*Attempts to persuade the reader* of the financial and commercial benefits of the proposal, by giving a breakdown of: • prices – a key [and often the key] consideration, together with a cost/payment plan • legal and commercial terms and conditions.
Management proposal/aspects of the proposal	*Aims to convince the reader* that the product (or project) can be delivered in the specified time and within the stated budget by detailing, for example: • production plans • delivery schedules • the company's organisation and management structures • research and development facilities • project engineers' CVs.
Appendices	*Support the technical argument* with supplementary data, for example test results and mathematical modelling

Since the Technical, Commercial, and Management sections tend to be the most substantial, they may merit separate volumes. The executive summary could be one of the shortest sections, unusually reaching 10 or 20 pages in length, but more usually comprising one or two. As is discussed later, the summary is one of the most significant parts of the proposal, because it may prove to be one of the most influential sections: chief decision-makers for the customer may base their decisions on the executive summary whether or not to short-list a proposal.

The table of contents, management section, commercial section, and appendices receive scant treatment in this book, because they have little bearing on the communication activities of engineers, or are not considered by them to be particularly central or problematic. The table of contents is automatically generated by the word-processor, and compiled by the technical author during the final stages of writing, with little attention paid to it by the engineers themselves; the management and commercial sections are usually written by colleagues in the Commercial or Sales department of the company, some of whom may have been engineers, or have an engineering background; and the appendices of an engineering proposal are no different from appendices to be found in any document, being optional, and their inclusion predicated upon need. In large bid projects, the appendices may be placed in separately bound volumes, together with supporting information, for example test results, query lists (comprising lists of specific questions requiring answers), or compliance matrices.

Traceability: a brief word on referential and administrative functions

Let us consider briefly a usually taken-for-granted aspect of the proposal document, that is, the referential and administrative function of information displayed on the front cover, since this relates to other reading and writing processes in engineering companies. Usually the cover specifies the proposal (name and proposal reference number), a bid reference number (issued by the customer), the date of printing, the company name (usually the prime contractor), and the key personnel involved in compiling the bid. Later discussion of particular proposal covers reveals other information which may be included, for example the names of subcontractors, logos, and visuals. The most obvious function of the cover is referential and nominative, in that it provides the reader at a glance with the reference number and name of the product, project, or both. In organisations where vast stores of documents are held, they have legal standing and security ratings, and so this feature is important to the smooth operation of working and writing practices of the engineers. The name and number of the proposal helps the customer with administering the reading of it, which is no small matter when several sets of proposals are submitted at the same time for different projects. The name and number are even more important for the administrative purposes of proposing companies, by aiding future location and reference, especially if it is the winning bid. As mentioned earlier, the technical proposal may form the basis of the design development of the product. The document may be referred to at a later date, and so it must be clearly positioned and easily traceable in the vast collection of documentation which will inevitably accumulate (these days, electronically) about the product. Engineering companies have a public duty of care to have a fast and accurate document and data retrieval system, and are accustomed to allocating time and resourcing to it. They are mindful these days of the practices expected of them, which in engineer-'speak' are referred to as 'configuration management practices', as detailed in ISO10007. Developing electronic data classification and storage systems has been a major cost burden in recent years, although they have had a better track record than other (mainly government) agencies and organisations in this respect.

The names of those responsible for the contents of the proposal, usually engineers, are placed on the title page or cover for three inter-linking reasons: reference, traceability, and accountability. These engineers would be considered accountable for the contents, and enquiries about the proposal (or product) would be addressed to them. Such enquiries may arise after some (distant) time in the future, possibly 10 or 15 years after the proposal is submitted, when engineers involved in the design of the product may no longer be working for the company. In such cases, knowledge about the product held in such documents is passed on to inheritor engineers as a sort of design-knowledge legacy. It is not uncommon for situations to arise where engineers need to know who 'holds' knowledge about particular products, since they may be the repositories of experience and wisdom. They are able to answer questions about decisions that were taken about the design, or particular problems which may have arisen during the development and use of the product.

7.3.3   The front cover: engineers' use of visuals

Highlighting ingenuity of design

From the 1990s, it has become more common to produce proposals with visuals on the cover, and elsewhere. There are various reasons for this, mainly, developments in desktop publishing, pressure from marketing colleagues, and engineers themselves overcoming their natural aversion to overt 'selling' ploys and embellishment. More recent team working has involved engineers more in marketing and selling the product and, as a corollary of this, in describing the product for the customer in proposals. Visuals are chosen with care by engineers leading bid projects, and their control of appearances can be absolute, as the following example shows. The proposal cover in Figure 7.4 concerns a silicon gyroscope (SG), a device for sensing changes in spatial orientation. In the case of the SG, the engineers

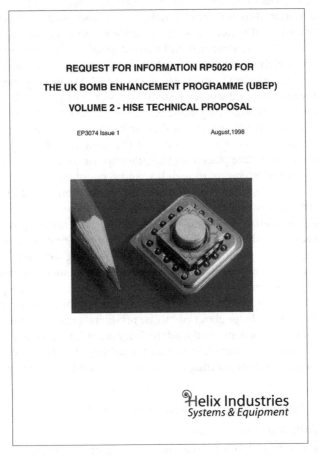

Figure 7.4　A front cover with the pencil picture

were noticeably proud of the engineering and innovative design of their product, and their success in designing a product with the dimensions of a pencil tip. It may look unremarkable, or even unattractive to the uninitiated, but to the engineers it was a work of art.

In this particular proposal, the engineer in charge (team leader) of the project had firm views about the picture he wanted to see on the front cover, rejecting others submitted to him by colleagues in the Sales Department, who were not engineers in this particular instance. In their eyes, this was a picture of what appeared to be an antiquated space ship from the first Star Wars film (in fact, the SG), juxtaposed with a pencil tip. The salesmen thought the picture rather crude and mundane and (in)famously referred to it as 'the pencil picture'. It galled them every time it was featured on proposal covers, as it often was. The team leader insisted on retaining this photograph in spite of their entreaties to replace it with, to them, a visually more attractive picture. He, on the other hand, was adamant in believing that the picture summed up the essence of the ingenuity of the product being offered in the proposal. He wanted to bring to the reader the realisation that the idea of the gyroscope as a spinning object was outdated, and that its construction no longer comprised moving parts. One of the outstanding features of the design was miniaturisation: it was an electrical component engineered to be no bigger than the tip of a pencil, with soldered leads to enable it to be connected to other components, and manufactured like a micro-chip. He believed the picture conveyed this, and had no doubt that the intended reader of the proposal, his counterparts in another company, would appreciate the significance of the juxtaposition of the component with a pencil tip, as would non-specialist readers. For the latter, he had included an explanation of the product in an appendix of the proposal, specially composed to 'educate' them about the design.

Intimations of an ideological stance

Engineers understand the usefulness of visuals in making proposals more effective and more attractive, and ultimately in persuading readers to recognise the superiority of their product. There is nothing remarkable in this observation: it is well known that pictures and all sorts of graphic illustrations, well chosen, may enhance a document and communicate a message more effectively. In engineering, though, the distinctive graphic genre that engineers favour is worth more than a cursory glance. Engineers' choice of visuals reflects the culture of the engineering discourse community. Myers (1990) and Kress and van Leeuwen (1996) noted the particular nature of the use of illustrations, or visuals, in the sciences. Influenced by the Hallidayan systemic approach to analysis (Halliday 1978 and 1994), Kress and van Leeuwen attempt a portrayal of the 'grammar' of visual images in a Hallidayan sense, stating that their book is about signmaking (1996: 5). It covers signifiers, which they refer to as 'forms' (they give as examples: colour, perspective, and line), and signifieds, which they explain as being the forms used to signify meanings. Their summing up of those involved in choosing images, and those who will interpret the images, encapsulates the situation of visuals in the context of engineering proposals:

> Interactive participants are therefore real people who produce and make sense of images in the context of social institutions which, to different degrees and in different ways, regulate what may be 'said' with images, and how it should be said, and how images should be interpreted. (1996: 119)

They also confirm the idea that the visuals used in proposals represent the ideological standpoint of the engineers, in that they project a certain image of themselves and the culture of their company through them. It could be said that they convey, through their selection of graphs, pictures, colours, and graphic designs, the way in which they would like to be seen. Although engineers like to think they are writing objectively and unemotionally, it is clear that there is a mismatch between this perception and actual fact. It is eminently probable that proposals are a representation of stance, for, as Kress and van Leeuwen point out:

> the apparently neutral, purely informative discourses of newspaper reporting, government publications, social science reports, and so on, may in fact convey ideological attitudes just as much as discourses which more explicitly editorialize or propagandize.... (1996: 12)

The pictures used on proposal covers have the purpose of communicating the ethos of the company, or providing pictorial contexts in which the product would be used. Readers can visualise the product, 'see' it in action, or project themselves into scenarios in which the product would be used. The picture of the SG used on the cover above would be regarded by Kress and van Leeuwen as having high modality, since it has as one of its purposes that of communicating 'scientific ideas or technological complexities to . . . non-initiates' (ibid: 169). They cite Myers' work on visual representations of scientific images for the non-specialist (Myers 1990), mentioning how these tend to be 'lavish, full-colour, and 'hyper-real', in contrast with the rather sparse drawings to be found in specialist scientific publications.

The examples of proposal covers, shown in Figures 7.5 and 7.6, depict the proposed 'solution' through visuals that aim to convey a dynamic context for the product, in the form of a dramatic pictorial narrative,

Figure 7.5   Proposal cover for a bespoke product intended for a destroyer

Figure 7.6   Proposal cover for a product designed for the Tornado aircraft

not unlike the dramatic portrayal of garter snakes in vivid pictures used in science writing, described by Myers (1990: 160, 167).

Influencing and impressing readers

Proposal covers used to comprise plain, monochromatic pages, with orthographic symbols (usually the title) as the focal point, accompanied on occasion by some kind of simple visual or graphic. They have evolved, concomitantly with developments in desktop publishing software, into the covers shown in Figures 7.5 and 7.6, depicting colourful images integrated into an overall complex graphic design. A technical author explained the idea of implementing his team's unabashed strategy to impress readers. He knew they would be scrutinising bids from different companies and explained it thus:

> We went from no images to these covers. We try to imagine three or four bids lying on the table, and aim to make ours noticeable. We want to make it stand out from the others and identifiable as our bid.

He explained that their motive for designing distinctive and attractive covers was probably the same as that for the design of a work of fiction about to be published. They wanted the design to make an impact, so that a reader would remark on the design, pick the proposal up, and be predisposed into thinking that it looked interesting. Of course, the old adage 'never judge a book by its cover' still holds true, if the story lying between the covers should prove disappointing. This could also apply to proposals, where the initial expectation inspired by the cover may not be fulfilled in the content of the proposal.

Another aim of the cover design is to make the proposal easily recognisable amongst all the other proposals and documentation the readers need to sift through. This has to be done without any knowledge about who their competitors are, usually, or what their competitors submit. A bid may not win, ultimately, but if the image conveyed by the visual image on the cover endures in readers' minds and renders the bid documents easily recognisable, one of the aims lying behind the use of visuals will have been fulfilled.

Engineers feel passionate about their designs and products. They attempt to convey an idea about the design on the proposal cover and in diagrams and pictures integrated in with the text. This is a straightforward enough matter if the 'solution' is outwardly visible and easily recognisable, like a helicopter, for example. It can be difficult, though, in the case of other products which are less visually evocative, romantic, or dramatic. The product may be nothing more than an unremarkable box, or consist of bits of equipment fitted in different parts of a ship or aircraft. It could be a piece of software, and therefore 'invisible'. Such products are difficult to depict visually in a way that would be interesting to readers. In such cases, the proposal cover may have a similar design to that shown in Figure 7.6, a proposal cover for an inertial navigation system intended for the Tornado aircraft, using global positioning by satellite. This complex system comprises various integrated items of software and hardware that are difficult to depict visually for a proposal cover. Instead, pictures of the aircraft, for which the product has been bespoke-designed, form the focus of the cover, with intimations of circuit-boards and other technical 'bits' as diffused images in the background. Similarly, the cover shown in Figure 7.5 shows the ship for which the product has been designed, the product being a large electro-optical tracking system. The story behind this proposal is a complicated one and worth retelling for the insights it provides into the cover design that emerged. Considerations of a political nature had a bearing on the design of the cover, since this proposal was affected by intricacies of inter-corporate relationships in the soliciting and submitting of bids. In this particular case, the customer (a government agency) had chosen an organisation to be the prime contractor responsible for the whole ship. This meant that the organisation was both overseeing and managing all the bids submitted and receiving proposals from its own subsidiaries. This placed the organisation in a sensitive (and difficult) position, in that it needed to be objective and above reproach in managing the whole bid process, and be seen to be so. It was a hugely complex process demanding the management of all the bids submitted for all the fittings, large and small, and all the sophisticated equipment to be designed, manufactured, and assembled, to produce ultimately a ship in its entirety.

The cover was designed to achieve several objectives. First, it had to downplay the fact that, as an engineer put it, 'we are bidding into ourselves', and so the engineers did not wish to flaunt this fact on the cover. They were second-guessing about the situation, and thought they might be disadvantaged by the fact that they were part of the prime contracting organisation. They did not wish to be seen to be favoured in any way in the proposal. For this reason, the government agency logo is larger and more eye-catching, and the logo of the submitting company is not prominently displayed, being, instead, juxtaposed with that of its subcontractor. An even more eye-catching logo, in colours not commonly found together on covers for engineering proposals (bright orange, yellow, and red), is placed in a prominent position on the cover to represent the new series of Type 45 destroyer, and to encapsulate the special working relationship between the government agency and its collaborating 'partner'. This logo shows the archer fish, which attacks its prey with jets of water, and which provided the original concept for the design of earlier versions of fire control systems produced by the company.

'Toys for the boys'

The final example of a proposal cover shown in Figure 7.7 is included as an example of a more generic type of product, in this particular case the gyroscope mentioned earlier. As can be seen, the picture encourages the idea of there being a wide range of applications for the gyroscope in a variety of craft. The picture is similar to ones used in brochures, advertising, and posters, and helps to convey a facet of the corporate image of the proposing company. It is an image the company wishes to project and one that decision-makers in various governments find appealing.

Figure 7.7 Proposal cover for product with wide application

Engineers agree that the image conveyed can sometimes be macho-oriented, evoking images reminiscent of old army comic books, or 'two-penny bloods' as they used to be called, read by young adolescent boys, and (much) older. Recurring themes are bravery, manliness, patriotism, camaraderie, fighting, shooting, explosions, and killing. Engineers readily acknowledge that the visuals used on proposal covers draw upon these themes to convey, as one engineer put it: 'smoke and flames and tanks and things', and another: 'toys for the boys', 'boys' in the latter quote referring to ex-service personnel working in government departments.

7.3.4 Proprietary page: laying claim to design rights

The proprietary page (or section) establishes the company's legal copyright for the contents of the proposal. The usual positioning of this on the page immediately following the title page is similar to that of copyright claims usually found in published books, although its length, around 150 words or more, and prominence may be quite different. Copyright statements in books are rarely more than 50 words long, appear in a minute typeface, and may be found inserted unobtrusively at the bottom of a page, or amidst information about the book's publishers, ISBN, key words, and so on. In company proposals, the proprietary statement is usually allocated a whole page, and is the responsibility of legal specialists in the commercial department. It is the only part of the proposal that engineers consider to be off-limits to them; they do not write it and are not allowed to change it.

Figure 7.8 is an example of a proprietary statement for a proposal, which stakes the company's claim to the ownership of the design.

Text as quasi-product

The inclusion of the proprietary statement and its prominent display demonstrates the strongly felt need by the company to protect its property. In this case the 'property' is the description of an engineering design, that is, the product (or 'solution', as engineers sometimes prefer to express it). The statement establishes the company's ownership of the design or design idea. There is another significant aspect to the proprietary statement: it reflects a proprietary attitude towards a product that exists even when it is still but a gleam in engineers' eyes. To them, the 'product' may exist even before the proposal has been submitted, and certainly at the time of writing a proposal. To explain the significance of this further, it may be the case that, initially, the product is manifest

Proprietary Statements

This is an unpublished work created in 2006, any copyright in which vests in Helix Industries Systems and Equipment (HISE). All rights reserved.

The information contained in this document is proprietary to HISE unless stated otherwise and is made available in confidence; it must not be used or disclosed without the express written permission of HISE. This document may not be copied in whole or in part in any form without the express written consent of HISE which may be given by contract.

This document contains trade secrets and/or sensitive commercial and/or financial information as of the date provided to the original recipient by HISE and is provided in confidence. Release of the information to any third party is prohibited without prior written consent from HISE. Public authorities are prohibited from releasing the information unless its release would not constitute an actionable breach of confidence. Public authorities should contact HISE to determine the current releasability of the information.

[5 USC 552(b)(4) and 18 USC 1905]/ [Sections 41 and 43 of the Freedom of Information Act 2000] are applicable.

UK Origin

Any enquiries relating to this document or its contents should be addressed in the first instance to:

John Smith
Helix Industries Systems and Equipment,
Bittaford Road,
Northway,
Northtown,
NT2 2EE

| Telephone: | | International: | |
| Fax: | | International: | |

Helix Industries Systems and Equipment and the ⊚Helix Industries *Systems & Equipment* wordmark are trademarks of Helix Industries Systems and Equipment.

Page 1

Figure 7.8   Proprietary statement: laying claim to design rights

as writing and the text and diagrams used to describe it. It is judged by the customer as it is expressed in proposal text, which he treats as a quasi-product. In other words, in the proposal, the idea about the design of the product is developed and established in textual rather than actual terms. As such, the text represents the product, and in fact becomes the de facto product until it rolls off the production-line and is delivered to the customer. The development of the product, and the expression of its design in text until its conversion from a textual into a physical product, is shown graphically in Chapter 3 (Figure 3.4).

Others have made similar observations. In her study of a professional community of tax accountants, Devitt views their texts as:

> a tax accountant's product, constituting and defining the accountant's work. . . . In return for their fees, its [the accounting firm's] clients receive texts – whether a tax return, a letter to the Internal Revenue Service (IRS), an opinion letter, or a verbal text over the phone . . . (Devitt 1991: 336, 338)

Observations about the text substituting for the product have also been made within the field of engineering:

> In many industries, such as aerospace or the automotive sector, there will be a considerable period when the product does not exist in a physical form. In the concept phase, for example, the product will be represented solely as a set of functional requirements. (Kidd 2001: 37)

7.3.5   Acronyms unlimited

A plethora of acronyms

Acronyms proliferate in corporate language, being a natural corollary of a dynamic, industrious work environment. They are commonly used and a distinctive feature of any large organisation. They form an integral part of engineers' texts, and the growth in their number seems unstoppable, with new acronyms being coined almost daily in some organisations. There may be someone whose job it is to keep track of acronym usage, so that a list of acronyms is compiled and stored in a dedicated acronyms binder or, in the case of companies with more advanced data systems, the acronyms are accessible online in computer databases. Lists are updated at regular intervals, because the development of new products, or the incorporation of new technology into products, spawns collections of new terms and coinings.

Acronyms are often the butt of ridicule, grumbling, and complaint. Any organisation has its pundits who, if asked about them, produce spontaneous diatribes about their use or, rather, misuse. Such reactions are understandable in the face of hindrances to reading that they may cause, and problems that arise out of misunderstandings by employees within organisations, let alone outsiders. The situation surrounding the (mis)use of acronyms has been commented on by Jones, who observes that this is where 'confusion abounds', and that even with a given list of abbreviations, no one organisation will use them in 'the same manner', nor will they be used with adequate definition of exactly 'what is being communicated'. He advises that it is essential to define clearly each acronym and abbreviation every time it is used (1995: 1.14–15).

There are two main reasons for the use of acronyms in written text. First, when one is used so often in engineers' speech for verbal economy, it being accepted that its referent is clear to interlocutors, it is automatically used to replace the orthographic form of the words it represents. An example of this would be the oft-used COTS (Commercial off-the-shelf), or DoD (Department of Defense). Second, acronyms are used for stylistic reasons, in an attempt to improve the language being produced and avoid repetitions of rather long expressions in the text. Certain expressions, like 'Application Specific Integrated Circuit', render text inelegant when used frequently, and may prevent it from being read easily. They may also be so long that substitute acronyms are useful for shortening sentences, which would otherwise be too long and unwieldy.

Sceptics may demur at this, but in a community of like-minded, like-qualified specialists, it is possible for the skilled use of acronyms to improve the readability of text. This, in turn, facilitates fast fluent reading, which is a key consideration for those with heavy reading loads, for example, proposal scrutineers. Naturally, overuse or unskilled use of acronyms could have the opposite effect, by impeding the reading process and irritating the reader, as often happens. Chronic misuse of acronyms has given them a bad name.

Presentation of acronyms

Acronyms receive special placing in engineering documents, usually near the front. A list of acronyms (or abbreviations), in alphabetic order, is usually allocated at least a whole page, no matter how short, and typically appears after the table of contents. Figure 7.9 shows an example of a relatively short list, taken from a proposal for the use of an inertial measurement unit in a product.

Conventions of use

The first mention of an acronym, similar to writing conventions in most formal business writing, occurs in brackets immediately after the word group it represents, for example 'Silicon Vibrating Structure Gyroscope

GLOSSARY

ALARM	Air Launched Anti Radar Missile
ASIC	Application Specific Integrated Circuit
HISE	Helix Industries Systems and Equipment
BIT	Built In Test
GDU	Guidance and Control Unit
NSRDS-DD	NIN Sparkwell Radar Defence Systems-Dynamic Division
GPS	Global Positioning System
IMU	Inertial Measurement Unit
INS	Inertial Navigation System
MoD	Ministry of Defence
OEM	Own Equipment Manufacturer
PCB	Printed Circuit Board
RFI	Request For Information
MPP	Makemoto Precision Products Ltd
SiGyros	Silicon Gyroscopes
UBEP	UK Bomb Enhancement Programme
UK	United Kingdom
VLSI	Very Large Scale Integration
VSG	Vibrating Structure Gyro

Figure 7.9   A glossary in a technical proposal

(SVSG)'. Thereafter, the acronym is used instead of the word group, as shown in the following extract, where the relevant sections are underlined for reading convenience:

> The Silicon Vibrating Structure Gyroscope (SVSG) is based on a planar resonating ring design. MATRIX has extensive patent coverage encompassing all aspects of the design and implementation. In particular, US Patent XXXXXXX relates to the use of the planar ring design with US Patent application YYYYYY, specifically covering the SVSG implementation including the drive and pick-off transducer mechanisation.

These are writing conventions encouraged by technical authors, although engineers may not always be aware of them or follow them consistently. In lengthy documents, or documents which they believe will be read in sections by separate people, for each new section, technical authors may repeat the practice of first using the expression in full, following it with an acronym in brackets, for the convenience of different reading teams.

Acronym categories – an example set of acronyms

In one company employing 350 engineers, the total number of different acronyms used in all types of documentation is currently around 4000.

Across a single document category (a corpus of 95 engineering proposals) a total of 1768 acronyms are used, yielding an average of around 18.5 glossary entries per proposal. This alone gives some indication of the reliance on acronyms in engineering documents.

Some acronyms have multiple entries because of different referents or other (slight) differences. In one document, for example, 'IP' may refer to 'Industrial Participation', and in another, it may refer to 'Initial Point'; 'AMS' may refer to 'Alenia Marconi Systems' or 'Aircraft Management System'; and 'MFC' may refer to 'Microsoft Foundation Class' or 'Multi-Function Console'. Of the 1768 acronyms found, 818 are individual, in that they have distinctly different referents. This total of 818 includes instances of an acronym being counted twice (or, occasionally, even three times), if it refers to two (or three) separate referents. So, if we consider the example of 'ALARM', which may refer to either 'Air Launched Radiation Missile' or 'Air-Launched Anti-Radar Missile', it is counted as being, and functioning as, two separate acronyms.

A closer examination of all the acronyms reveals a varied list, with the potential for sub-categorisation. A thumb-sketch overview shows that significant proportions refer to:

1. *Organisations*, for example NATO (North Atlantic Treaty Organisation), RNLF (Royal Netherlands Air Force), RSC (Raytheon Systems Company), SAGEM (Société d'Applications Générales d'Electricité et de Mécanique), MBDF (Matra BAe Dynamics – France).
2. *Whole systems: objects, products, vehicles, or equipment*, for example COPICS (Communication Oriented Production Information and Control System), AEU (Antenna Electronics Unit), ALARM (Air Launched Anti-Radar Missile), MBT (main battle tank), RPV (remotely piloted vehicle).
3. *Technical, scientific, or mathematical terms*, for example: CMT (Cadmium Mercury Telluride), RMS (root mean squared), SNR (signal to noise ratio), BCD (Binary Coded Decimal), LSB (least significant bit), MSB (most significant bit).
4. *User-focused location and orientation*, to reflect how the product may be used by the user, for example HU (head up) and HD (head down); they may consist of positional adjectives (high, low), and parts of the body (head, eye), for example ESL (Eye-safe laser), and LRU (line replaceable unit), which refers to components that may be easily detached (removed) and replaced by the user, who may be using equipment in the front-line, say.

Influences on acronym usage

Technical authors usually have the final say on the use of acronyms or abbreviations, and whether they will improve or ill-affect the readability of the text (or, less often, the comprehensibility of an oral present-ation). Clearly, as one author put it, they do not want to 'turn off' readers by peppering the text with unfamiliar acronyms. They are aware of how injudicious usage can ill-affect the readability of engineering documents. In the special case of technical proposals, however, they may have little choice but to use acronyms specified by the customer, who will have used them in the Customer Requirement. They may feel obliged to use them, in spite of their stylistic objections, not wanting to be seen to reject the customer's terminology. On the other hand, they also know that engineering language can be difficult for non-engineers to understand at times and that certain documents need to be written with the non-specialist reader in mind. In such situ-ations, it may be stylistically more appropriate to minimise the use of acronyms.

More commonly, however, readers are familiar with the terminology, and possibly because of this, some engineers assume their readers are broadly familiar with the acronyms used. In fact, there is a minority of engineers who believe, rightly or wrongly, that acronyms will impress the reader and, furthermore, that the reader actually expects to see them in texts and to hear them being used in engineering talk. Every discip-line has specialist vocabulary and jargon, and so there is some logic to this view: engineering groups (and sub-groups) use vocabulary and acronyms that mark them out as distinctive, and their use symbolises exclusivity of the group. There is, then, some point to the view that sees the use of acronyms as a necessary form of professional display, or a kind of professional 'showing-off', which is expected by fellow members of the group.

7.4 Oral presentations of proposals: a case study

Description of communication tasks within the engineering domain tend to focus on written communication. Engineers' concerns about communication at work are more often associated with written texts than spoken, and this concentration on the written word is reflected in this book. However, just as earlier chapters have shown design engin-eers to be concerned about texts relating to product development, it follows that they should be equally concerned about the way they talk about it. And they are, especially when presenting information about it to the customer. In fact, as earlier chapters have observed when describing engineers' opinions about communication, engineers rate oral communication tasks highly, for they appreciate the effectiveness of clear expression in both speech and writing (Huckin and Olsen 1983: 167). Chapter 4 refers to events requiring engineers to make formal oral presentations, which include those to colleagues, as in the case of briefings and training sessions, and to the customer.

Engineers do not make oral presentations for all proposals. Such presentations are common in competitions where the sums involved are large. Usually engineers present for larger proposals worth £5 million or more. This is a rough estimate, however, since what may be considered large to one company may be small to another larger company. Nonetheless, size considerations apart, there is an understanding in the industry that, in principle, proposal teams are willing to present a proposal to any customer who wants it, no matter how small the bid.

This section describes the events surrounding a particular oral presentation that a group of engineers had to make to the customer, in this case, members of a department at the Ministry of Defence. The engineers, drawn from a company in the United Kingdom and another European Union country, had submitted a joint bid for a contract to design, manufacture, install, and maintain a gun system for Royal Navy frigates. To the companies concerned, it was a big contract and important to win in order to secure employment stability for employees for years to come.

Only short-listed proposers usually have to make presentations, and on receiving the news that their bid had made the shortlist, the project leader organised presentation rehearsals in preparation for the event. Five engineers drawn from the two bidding companies formed the presentation team, three (the first three in the following list) from the company in the United Kingdom, the other two from a non-British company, referred to in the research journal entries as NORTHERN:

1. Michael, who was the bid leader, was regarded as articulate and experienced by his colleagues, who relied on him to help them write more suitably for the proposal in question. His scrutiny of all their written contributions to the proposal documents was explained by one engineer thus: 'He goes through our writing from the point of view of the customer.'
2. Colin, an electrical engineer, whose main role was to provide a technical description of particular aspects of the design, including the system architecture (both internal and external).

3. Alex, a mechanical engineer, who would be in charge of managing the project if the proposal succeeded. He was responsible for describing the proposed project schedule and customer/user support (referred to more usually as ILS, which stands for 'Integrated Logistics Support').

4. Jean-Paul, a mechanical engineer, whose task was also to provide a technical description of parts of the solution that his company would provide.

5. Heinrich, Michael's counterpart, responsible for marketing in his company, who would describe to the British customer the role played by his non-British company in the project. With just a week to prepare, the presentation team initially spent the first few days at their respective companies preparing PowerPoint slides and individual scripts for their own contributions. They used email, the telephone, and video-link for inter-country communication to keep in touch with their colleagues in the partner company, but practised in-country, arranging to meet the day before, for a run-through of the whole presentation together. The engineers found this was a time of concentrated, frenetic activity, when they worked closely together in preparations and rehearsals, in a similar way to that described by Ochs and Jacoby, who worked among physicists preparing conference talks (1997). Journal entries for the occasion now provide a description of the events leading up to the present-ation and the presentation itself. In-between the journal entries, that now follow, is a summary of the main comments made by those watching and evaluating the team's performance during the rehearsals.

Research journal entry: Rehearsals – a gruelling time is had by all

29 April: I remember Michael saying, when we talked about the criteria used for judging a proposal and when I suggested that price was probably the main factor (although in the past I'd thought it was compliance): 'If you've got what they want, the money isn't that important. If you haven't got what they want, the money is irrelevant anyway.'

I wonder if, judging by the way the presentation helped the other company to move from 4th to 1st, it is starting to sink in that these oral presentations may be a crucial part of the bid process, especially as the selectors/shortlisters seem to be using other criteria to help them select. They're certainly talking about this dramatic elevation quite a bit. It's making them very thoughtful. And the customer wants to meet the people who will be actually involved in the project if the proposal wins. This makes me think about the importance of having good reliable key people in place. Michael told me that the customer likes to see the actual people on the team who will be working on developing the design. So the presentation team on a large project like the last one just recently (CIRRUS) comprised Michael and Tim as the Marketing and Engineering spokesmen, Andy Slade, and two other 'boffins' [my term] who were the engineers working on the secret bits/core parts of the design. Also a guy from Longue Madison, who plugged his bit of the presentation into the middle of the PowerPoint presentation. Alex says you can tell where his starts: 'see the dense bit of text in the middle of all the "light" stuff'.

1 May: Alex has passed onto me their presentation on Power-Point to look at. The whole team is working intently to prepare their scripts, and Alex wants me to go over some of their sentences. They seem to be crowding out their slides with too many words.

3 May: Today, no sooner than I had got in and settled, Michael asked me if I'd like to sit in on their practice run-through of the presentation they'll be giving to the POC, as they refer to it, i.e. the prime contractor. What an opportunity! I joined the audience of only six, but they were all senior managers, a couple of whom didn't pull their punches. It lasted nearly three hours, during which Michael, Colin and Alex were standing most of the time. They weren't obviously agitated, but I think they must have been quite nervous. The comments they were getting weren't personal exactly, but they were pretty direct, like Alex was too wooden (which he was, almost zombie-like), Colin didn't sound at all sure of his material, and Michael seemed to be just plain distracted. I think some academics wouldn't be able to tolerate such feedback, but these three just soldiered on. When it came to an end, the managers all rose as one, nodded to the presenters, and trooped out, duty done. They are obviously used to providing this kind of feedback on oral output, just as they do for written output. Focused, candid, direct, but somehow impersonal and objective – all part of their work procedures, I suppose.

Continued

In the afternoon, I watched Alex go through his presentation again, with Gerald Maybin, Anthony Monk, and Cyril James as the audience this time. It lasted nearly three hours, and I started to find it gruelling. Goodness knows how Alex must have felt. His bit went on for two hours plus, and then Colin ran through the other slides for himself and the NORTHERN guys [NORTHERN is the collaborating company]. Still, Alex managed to appear a bit more animated; just a bit. It's funny, in between his performance pieces, when he was making asides and talking about his performance, his voice was much more interesting to listen to.

Later I typed up feedback notes on my impressions (including one or two comments made by their bosses), which they were impressed with apparently. They each came and talked to me about them, and seemed surprised by some of the observations. Maybe no one has given them this kind of (linguistic) feedback before. Alex said he was still talking to me in spite of my comments; wants me to watch him again tomorrow. Colin said he'd have it out with me tomorrow. All friendly banter, . . . before the storm. They're talking of taking me with them to the real event. There are a few questions about security, but Michael is going to see if he can pull it off.

7.4.1 Feedback on oral presentation

This is an account of the comments the presentation team received after their practice run-through. It includes a selection of the comments that were written up for the proposing team to consider before the actual presentation four days later. The comments reflect the overall impression of those watching the first rehearsal that the team had a good solution to present, but needed to improve their presentation of it in various respects.

Positive feedback

1. The main story seemed to be about the strength of the proposal team.
2. Effective team working.
3. The team came across as being committed to the solution and convinced of its efficacy.
4. Team members were experienced, knowledgeable, and knew the project inside-out.

5. Team members came across as involved and concerned, and talked fluently about the proposal.
6. It was a collaborative/cooperative proposal, a genuine working alliance with NORTHERN.
7. Ingenious technical solution.
8. Companies' commitment to the project.
9. Carefully thought through product schedule.
10. The engineering solution comprises tried and tested software and components. This meant the companies don't anticipate any problems with producing the product.
11. The solution was presented from the users' perspective at one stage, showing an understanding of the user's needs.
12. A broader view of the customer's needs was provided, showing the team had thought about features he hadn't asked for but may have wanted. They were also forward looking, giving a long-term view.

Negative feedback

The audience acknowledged the special difficulties associated with presenting to colleagues, and that the team's presentation style would probably rise to the occasion for the real event.

1. *Information about the project team* needed to be expanded. The audience wanted to know who the team members were, how long they had been with the company, and the work they did. More interesting and detailed information needed to be given to enable the customer to get the measure of the people working on the project. The introductions (and self-introductions) were hurried and rather glossed over.
2. *Glibness* is easily recognised and disliked by the customer. Catch phrases Michael used [responsible for marketing], like 'the team will deliver' and 'we are the best', etc., are unconvincing and a potential turn-off.
3. *More positive language* needs to be used. Statements need to be couched in more positive terms. Often, sentences began with a negative clause or a problem, which the audience did not always associate easily with the solution mentioned in the next clause or sentence, for example, Alex said 'We've been having a few problems with X, but we'll be sorting it out whenever . . .'. There were quite a few negative constructions like this, so that the language came across as more 'problem' than 'solution' at times. There was a tinge of this kind of negative

colouring in much of the language used, which belied the confidence the team genuinely had in their solution.

4. *More customer-focused language* was needed. There were times when the customer seemed peripheral to the engineers' concerns. There wasn't enough inclusion of the customer in the language. The engineers were presumably talking to the audience about what the audience's needs were, but kept referring to them as 'the POC', as in 'If there's anything the POC requires, we can...'. The head of procurement, who was in the audience, asked if there were reasons for not using 'you' or some less distant term.

It was also suggested the engineers think about rephrasing utterances like the following to make them more customer- and less presenter-centred: 'We'll have this...; we'll have that...; In July, we'll also bring in XXX...; We're going into a developing scheme with such and such...; We'll be provided with...by NORTHERN...'; 'We needed X...; etc. Colin in particular used these frequently when describing aspects of product development.

5. *Avoid being too self-deprecating* to the extent that benefits of the 'solution' become obscure or are downplayed. There were instances when the engineers seemed uneasy about touting the benefits of the solution, as though they saw it as boasting and were uncomfortable about it. For example, Colin started each new point (about the benefits of the solution) by saying the opposite to what he really wanted to mean. So, if he was pleased with the mature stage at which a component had already been developed and used, he said something to indicate the opposite. His language set up the wrong expectation in listeners, who needed to listen on carefully to get the gist of what he was saying. Colin was proud of the 'box of tricks' they had designed, but introduced it unenthusiastically with the words: 'Basically, it's just a box full of cards.' In fact, all the presenters used 'just' in a way that tended to devalue the message, for example: 'This is just a slide of the ship we used for trials'; I'm just going to talk about X'; 'This is just the software we're going to use.'

6. *Eye contact* with the audience was not very good. Michael's and Alex's gaze tended to be fixed, hovering vaguely away from or above the audience for most of the presentation. This had an excluding effect on the audience.

7. *Voices* were generally clear and easy to hear and understand. However, voice 'tune' (intonation, musicality) needed to be improved. Michael and Alex sounded rather flat and monotonous, with their voices tailing off at the end of sentences or, and this was more important,

went into a low, downward-falling note (referred to as a proclaiming tone in linguistics) (Brazil 1985). This could have a negative effect on listeners, who may be inclined to 'close down' in their minds the idea that the team is just beginning to develop, because of the falling pitch of their voices. It was noticeable when Alex in particular switched from his formal, rather dead-pan presentation voice to quick asides to colleagues. These were more natural sounding and better to listen to.

8. *Body movements* were either too controlled or distracting. Colin and Alex hardly moved a muscle, and seemed overly constrained by the controls for the display equipment. They seemed stiff and uncomfortable. Michael kept fiddling with his hair, in an absent-minded way. This distracted from what he was saying.

9. *Pausing* needed to be used to good effect. All the presenters tended to rush their words and run sentences together without seeming to draw breath. They needed to pace their language more effectively and pause at times to emphasise points, and to allow audience mulling time.

10. *Negative points* should be avoided. Negative associations of Component X needed to be downplayed. The significance of Component X as part of the, to use Martin's words, 'baggage' needed to be portrayed differently, because it came across exactly as that in the presentation, that is, negative baggage. Component X had caused problems for the customer (and the company) several years before, but in the end was developed into a worthy piece of equipment. It was mentioned in association with the word 'problem' several times during the presentation. If it had to be mentioned, the team were advised to use language to portray it in a more positive light, or to reconsider having to mention it at all. It was all very well being honest about past problems, but there was little point of mentioning them if they were no longer relevant.

Research journal entry: Rising to the occasion

7 May: Too much happening. Saw the presentation. London. Just yards from the headquarters. Tight security: I even had to be escorted to the toilet by a male security guard. Fascinating experience to be there during the actual presentation and the grilling question-time afterwards. It opened my eyes to see 'the customer' in action,

Continued

all 9 of them, ranging from a few young lads (all inscrutable, absolutely pan-faced) to some bewhiskered RN type, who was probing, pernickety, and only interested in how well the guns would fire in wartime... 'to hell with peacetime' he yelled. All seated round this massive oval wooden table. On the one hand, bright young science graduates, no doubt plucked straight from university who didn't say much, but what little they did say was clever, and on the other, a couple of old salts that have been on battleships and frigates from the age of sixteen (or so it seemed. Also very smart). They weren't shy of showing their interest in the equipment, and fired questions in quick succession. The team took to it like ducks to water. The presentation was very smooth and all came together like magic. Jean-Paul and Heinrich dovetailed their parts seamlessly. None of the awkwardness of the rehearsals. Everyone was smiling and seemed relaxed, though they told me later they were not!

[*Author's note*: Research journal entries in Chapter 6 about the LAWD bid explain what happened after this.]

7.5 Conclusion

This chapter has shown engineers' interest in any presentational aspect of the product. Their concern about the minutiae of design is extended to anything that is written (or said) about it that may be made public, especially to potential customers. The proprietary attitude that companies have towards their product is matched by the sense of ownership that engineers have about their designs. Their professional pride in their work is a reflection of this, and it is axiomatic to them that it be presented in the best possible light. Of course, this is important in the case of proposals, where success or failure may hinge on how effectively this is done. Engineers, and those who support them in project teams, know this full well, which is why so much time, energy, and resources are invested in presentations both written and oral. The next two chapters examine more substantial aspects of proposals, looking beyond cosmetics to their content to see how engineers attempt to persuade the customer, through particular information about the product and the company developing it.

8
Engineering Proposals: Discourse and Information Structure

8.1 Introduction

Proposal writing came to prominence in the late twentieth-century as a significant revenue-raising activity across most sectors: commercial, academic, and charity. Myers describes proposals as being from a practical standpoint 'the most basic form of scientific writing' (1990: 41) in his discussion of research biologists. However, whatever the domain, proposal writers rarely find proposals easy to write and it is rare that they are happy with what they produce. Engineers are no different. They and their managers are generally dissatisfied with the proposals they write and ask to be provided with proposal writing models, guidelines, and books that will tell them how to write them.

Linguists, on the other hand, may baulk at requests for writing models if they are dyed-in-the-wool descriptivists, and may be reluctant to accede to such requests. This chapter is a compromise. It examines the discourse structure of proposals, referring to work that has already been done in the area, and attempts to identify all relevant discourse functions. It then uses these functions as a basis for devising the structure for a generic proposal. Primarily, this analysis attempts to be useful to engineers, although it also attempts to demonstrate the efficacy of an applied linguistic approach to real texts in the workplace.

8.1.1 Proposals persuade

Most engineers would consider it a simple question to be asked about the structure of a proposal, and would reel off on their fingers the main section headings to be found in most of the proposals held in the company's database. Similarly, they would have no doubt about the ultimate purpose of the proposals, and would probably say these words, which were said by an engineer, or something very similar:

> Proposals are persuading documents. They have to persuade the Customer that ours is the best solution.

This is simply put, and goes to the nub of proposals but, when questioned, engineers find it difficult to explain how they try to be persuasive. It is, after all, a slippery concept and hard to pin down, as Chapter 6 has shown. However, it is the linguist's job to take the notion of 'persuasiveness' and try to identify aspects of the proposal that are persuasive, be they organisational, visual, propositional, or stylistic. (The use of 'propositional' is similar to Searle's use of the term (1969: 29) and concerns information content.) The starting point for this section, then, is that, whatever the stimulus, and without exception, proposals set out to persuade. They are written in order to convince the customer of the efficacy of the proposed 'solution', that is, the 'what' aspect of what is being proposed, which could be a product with physical properties, a set of documents, a piece of software, or a procedure.

This view of proposal writing as strategic and reflecting engineers' writing motives has been informed by the work of Swales and Feak (1994), in particular, their approach to teaching academic writing to graduate students, which arises out of earlier work on genre analysis (Swales 1981, 1990) that is particularly relevant to proposals. A recurring point made throughout their work is that academic writing is rhetorical, and that they see rhetorical writing as strategic writing:

> All of us, as academic writers and whatever our backgrounds, are engaged with thinking about our readers' likely expectations and reactions, with deciding on what to say – and what not to say – about our data, and with organising our texts in ways that meet local conventions . . . (Swales and Feak 1994: 3)

It follows, then, that when engineers are engaged in proposal writing, they are similar to other writers, like the post-graduate academic writer, in that they want to achieve an outcome, and persuade the reader into a certain type of behaviour; hence the aptness of the terms 'strategic writing' and 'writer motive'. It is the combined aspects of motive and persuasion in proposal writing that distinguishes it from other types of writing produced by the engineer when designing a product.

8.1.2 'Selling point' versus 'benefit'

Chapter 1 discusses the engineers' usage of the word 'product' and the distinction they make between it and 'solution'. As a further illustration of engineers' sensitivity to semantic nuance, let us consider the significance of the terms 'selling point' and 'benefit' with regard to the language of technical proposals. Engineers tend to see the terms as context-differentiated synonyms, since they use both in discussions, although context of use may determine whether 'selling point' is used, rather than 'benefit'. If engineers are more 'customer-focused', for example, and wish to reflect this in their language, 'benefit' is the preferred term. However, strictly speaking, from the company's perspective, these would be considered 'selling points', which conveys a more profit-oriented perspective. There is a subtle difference between the two, but it would take a rather crass engineer to use 'selling point' when making a presentation to the customer. Instead, aspects of the solution are referred to as 'benefits', since this conveys a more customer-oriented standpoint. Engineers are sensitive to the semantic differences between the two, probably because they are pulled in opposing directions. There is a tension between their wish to cater to the customer's desires (and to see the solution from the customer's perspective) and their need to make a profit (out of the customer).

8.2 Guidance on proposal writing: a historical perspective

There is a perception among engineers, those who teach them, and those who help them to write proposals (the technical authors), that there is a lack of useful guidance on proposal writing. This section provides a survey of work already published, in order to bring together received wisdom on proposals and to shed more light on this complex document. The situation is somewhat similar to the one described in Chapter 2, concerning guidance on engineering procedures. However, to report, without some qualification, that a dearth of information exists on proposal writing would be paying a disservice to the considerable and informed literature that exists in the area of English for Science and Technology (EST) and English for Special Purposes (ESP). The fact remains, nonetheless, that this is a neglected aspect of engineers' writing, and always has been. If we consider the substantial work on EST that took place in the middle decades of the last century, mainly in America, it would appear that more attention was paid to proposal writing before the 1980s than afterwards (Souther 1954, Marder 1960,

Hicks 1961, Weisman 1962, Pauley 1973, Souther and White 1977, Fear 1977, Houp and Pearsall 1980).

8.2.1 Up to 1980: the popularity of report writing

Much of the work of this time arose out of the distinctive applied analytical practices and close working relationship between applied linguists in academia and industry, as exemplified by the work of those at the University of Washington (Souther and White 1977). Even so, before the 1980s, proposal writing was rarely mentioned. Technical report writing was the major concern around this time, and these books were written primarily as text books to be used on ESP/EST courses for students in universities and other tertiary institutions, and in-company training courses. Across all the works, the emphasis is generally on the writing of manuals, feasibility studies and the like, technical articles and/or papers, letters, and memoranda. In acknowledging that engineers write a range of documents, Hicks, for example, makes fleeting references to proposals but uses the term 'proposal' to refer to recommendations made in technical reports (1961: 141). He also uses it informally to mean 'suggestion'.

In the two books which explain proposals in depth, proposal writing is subsumed under 'report writing' or 'technical reporting': Pauley categorises them as a type of formal report (1973: 163), and Houp and Pearsall designate proposal writing as one of the applications of technical reporting. They state that proposals are a kind of technical report, and deal with them in a section that includes progress reports, feasibility reports, and correspondence (1980: v, 267). Accordingly, Houp and Pearsall deal with proposals as sort of quasi-reports, and anyone who has researched or taught report writing will be familiar with their prescription: 'Introduction-Body-Solution-Attachments', an oft-quoted outline structure for reports. They suggest writers should follow this outline plan in proposal writing (ibid.: 344), adding, as a rider, that such an outline applies to small-scale proposals written by one person, and implying that what they refer to as 'mammoth' proposals would be structured differently:

> You should understand, however, that the paperwork for mammoth proposals (investigations in the millions or billions of dollars) may fill a five-foot shelf. (Houp and Pearsall 1980: 345)

This is the only reference to large proposals, since their chapter on proposal writing deals mainly with unsolicited 'short-form letter

• Reference to earlier association	• Time and work schedule
• Subject and purpose	• Facilities available
• Definition of the problem	• Previous experience
• Immediate background of the problem	• Personnel and their qualifications
• Need for solution of the problem	• References
• Benefits that will come from solution	• Likelihood of success
• Feasibility of solution	• Products of the project
• Scope	• Cost and method of payment
• Methods to be used	• Descriptive and advertising literature
• Task breakdown	• Urge to action

Figure 8.1 Houp and Pearsall's list of items for proposals (1980: 345)

proposals', which are uncommon these days. They discuss the sort of information a proposal could contain, which they list as separate topics, with the advice that only those 'items' that are pertinent should be included. The full list is provided in Figure 8.1.

They suggest to proposal writers that these 'items' be used, or combined, as headings, implying that the order of appearance of the items could be useful to follow in an actual proposal. The advice is generally expressed, relating more to writing purpose and strategy than to specific language or textual features, as illustrated by their discussion of 'Definition of the Problem':

Depending upon the scale of the proposal, you should spend from a paragraph to several pages in defining, locating, and describing the problem you propose to solve. By this means you may 'shock' your intended client into sudden and full awareness that a problem really does exist. However, you should guard against overstatement and overdramatization, because the techniques can boomerang. (Houp and Pearsall, 1980: 347)

Much of Houp and Pearsall's chapter is devoted to explaining each 'item' in this way, offering advice and using narratives for analogies to be drawn, or case-study-type explanations, to help the reader. Information about how one would actually compose defining, locating, or describing text is provided in an earlier chapter on technical exposition, which deals with such writing topics as exemplification, definition, classification, comparison and contrast, and so on. The essentially persuasive intent of proposals is mentioned briefly, when they state:

... we can safely assert that a proposal is designed to discharge two salesmanship functions:

1. To get a proposal accepted.
2. To get you (or your company) accepted to perform work (Houp and Pearsall 1980: 344).

Engineers need to develop expertise in writing persuasive proposals, and they need explicit advice on how to do this, for today, as in the 1980s, engineers remain ambivalent about writing persuasively, regarding it as a kind of writing they would rather not be associated with. This issue is explored in Chapter 6.

Pauley draws a different distinction between proposals: rather than seeing proposals as either solicited or unsolicited (the distinction made by Houp and Pearsall), Pauley sees a clear division between what he calls 'interfirm' and 'intrafirm' proposals, the difference being, in his interpretation, that interfirm proposals are competitive between bidding companies and submitted as part of a tender process, whereas intrafirm proposals are unsolicited proposals made within a company, usually from an employee in a subordinate position to his 'superior' or manager. Of the little information that is available on proposals, Pauley's work on interfirm proposals most accurately reflects the current situation with competitive proposals in modern engineering companies. However, like his peers', his treatment of them is scant, with most of his discussion being devoted to intrafirm proposals, which he portrays as non-competitive, written for an internal audience within a company. However, these days, intrafirm proposals are more often than not competitive, as in the case of bidding for research and development funding by the company, for which colleagues may also be competing.

Nonetheless, Pauley attempts to provide a framework of some use to proposal writers generally, which would be applicable to both inter- (i.e., competitive) and intrafirm (non-competitive) proposals. He provides an outline, two aspects of which would be familiar to current proposal writers (1973: 166). First, in the 'body of the report', Pauley uses the combative phrase, 'Plan of the Attack', to label a part of the technical section. Such a label reflects the strategic approach needed when drafting the technical part of the proposal, and Pauley's choice of words conveys this. These days, though, engineers may take issue with this aggression-loaded tag, preferring to see their proposals as tailored to the customer's needs. In their documents they are proposing a technical solution to the customer's problem and want to win the customer over to their side, not antagonise him.

The second aspect relates to Pauley's identification of three major subsections in the 'body': Technical, Management, and Cost. This tripartite division continues to the present day as major and distinct parts of the proposal, which are sometimes submitted as separate volumes (see Chapter 7 for the structure of modern solicited proposals). Pauley's overview of the interfirm proposal uses as contextual illustration the Pentagon's invitation to companies to bid for work on the B-1 super-sonic bomber (ibid.: 163). He provides a concise gloss of the function of the main parts of the proposal, as can be seen by the following:

> *Technical.* A proposal's technical section begins by stating the problem to be solved. This seems unnecessary, but the firm must clearly demonstrate that it understands what the solicitor expects. Then, the firm describes its approach to the problem and presents a design for the product if one is needed. Sometimes, the firm offers alternative methods for solving the problem and invites the solicitor to select one. (Pauley 1973: 165)

However, as is the case with Houp and Pearsall's work, this is the extent of the information provided on how to structure proposals.

8.2.2   Post-1980: proposals receive scant attention

If proposals were not pre-eminent in pre-1980s works, they have been positively neglected in the decades following. A trawl through more recent writing in the field of applied linguistics or ESP yields little specifically on proposals. A search through numerous writing guides and books on communication skills of the self-help 'close-that-sale', or 'write-that-winning-proposal' variety, yields poor pickings. Such public-ations, including those intended specifically for engineers and technical authors, generally pay scant attention to proposals, and are not repres-entative of the time and effort spent on proposals by engineers at work today. There are, however, a few notable exceptions, for example Freed (1987) and Ellis (1997), and, in the non-academic professional field, technical authors and engineers have found the ideas put forward by Stross (1990) and Covey (1997) useful.

Ellis starts by defining the proposal as being essentially 'a selling document', citing the MoD as a major customer (1997: 166), in much the same way as Pauley, mentioned above, names the Pentagon in the opening section to his chapter on proposals as being a significant instigator of invitations to tender. Clearly, governments continue to be a major source of business for engineering designs. In Ellis's and Stross's work, little attention is paid to textual or language aspects, their main concern being the human motivation behind proposal writing and concomitant aspects of writing strategy needing to be adopted. In both cases this involves examining the sort of information that should be included in order to follow the strategy. Their guidelines are usually generally stated, or have general applicability, with the understanding that persuasion is the primary purpose. Persuasion permeates all their discussion of proposals.

This sort of treatment of proposals can be seen to be sociologic-ally determined, with the primary emphasis on interaction (Brown and Yule 1983: 228), since readers are asked to first consider relationships between proposal writers and customers, and to use this to inform their writing strategy. Ellis, for example, provides a list of bulleted topics to be included in a proposal, which includes costs, quality control systems, and ability to keep to deadlines. He follows the list with advice that emphasises the interactional aspect:

> There will be many other questions, but unless the organization receives reassuring news on these then it is not much point going further with the relationship. Naturally a proposal has the best chance of 'winning' if it closely matches the customer's needs... If for any reasons the proposal that you write (either individually or as part of a team) is *not compliant* then the reasons for this must be clearly spelled out. You must be able to persuade the reader that your reasons are acceptable and can be justified. (Ellis 1997: 167)

Note his informal and direct style of writing, which engineers seem to find appealing, and the mention of issues that they can relate to, for example, cost, compliance, and delivery deadlines. These are the sort of issues that may typically arise, and showing clear relevance to authentic work problems is an approach commonly followed in popular self-help books. Without doubt, the target readers for these books will find much that is relevant to proposal writing, although it is usually business-related aspects which are dealt with, rather than language. For example, Ellis suggests performing a type of SWOT analysis, where the writers answer questions relating to their commercial strengths, weaknesses, opportunities, and threats, in order to decide on future business and, in the case of proposals, the writing strategy to be followed. He provides proposal writers with what he refers to as 'techniques' for preparing 'such a major document as a proposal', and suggests they follow the project management outline he provides (ibid.: 168).

He emphasises that proposals need to persuade and, as part of this procedural framework, suggests that a question-posing approach be followed to persuade the customer. The questions are intended to serve as a focus for writing, by helping proposal writers establish broad topic areas to write about:

Why us?

What have we got going for us to help win this bid?

Where do we score? How can we play on our assets?

How much will the customer know about us?

Do we need to correct any misinformation that our would-be customer may have?

Why not us?

What are our deficits? How can we circumvent these? How can we turn a seeming deficit into a strength or opportunity (i.e. we are a small organization, but we can be flexible and rapid in meeting your requirements; we don't have any cumbersome bureaucracy to slow us down!).

Why another?

Who are our likely competitors? Can we assess their relative strengths against us? How can we gear our proposal so as to minimize their relative strengths and maximize ours? (Ellis 1997: 168)

The questions, which any proposal team would find relevant, are included here to show their essentially general, context-setting, and purpose-establishing nature. Once the questions are answered, however, there still remains the problem of structuring and writing the proposal document, and it is the more specific advice for this which is obviously lacking from this book and others. Should the reader wish to discover more about writing, he is referred to 'writing techniques' in an earlier chapter on report writing, which deals with such topics as readership, terms of reference, overcoming writer's block, outlining, project planning, and data collection. This is really quite similar to the approaches adopted by Pauley (1973) and Souther and White (1977), discussed earlier. It seems that the literature on proposals generally omits aspects that engineers have more difficulty with, that is, information content, discourse structure, and language. It is these that we shall study in later sections, although first we need to examine exactly what engineering proposals set out to achieve. The next section describes the fundamental aims of these texts or, to use an applied linguistic term, their 'discourse functions'.

8.3 A hierarchy of discourse functions

From observations made on proposals in earlier chapters, it is possible to suggest a hierarchy of discourse functions relating to proposal documents.

8.3.1 Macro-level discourse functions

At the highest level, the proposal is a:

1. *response to a (potential) customer.* The proposal is a response to a textual stimulus, usually written, which can be an expression of need, expressed in a formally stated Requirement, and/or an enquiry for information (see Chapters 5 and 6).
2. *description of the 'solution' or product.* For engineers, this is the most significant function of the proposal. Again, it needs to be performed with a view to persuading the customer, and achieved through the performance of three sub-functions:

    i) definition of the product (or solution), and definition of related terms.
    ii) explanation of what it is and what it is capable of doing.
    iii) proof or demonstration of the extent to which the product gives the customer what he wants, that is, how far it fulfils the customer's requirements (referred to as compliance).

3. *means of persuasion.* Proposal writing involves engineers in a complex problem-solving exercise, the result of which must be a solution the customer likes. They need to convince the customer that their solution is the best response or, if the customer has a problem, that they will solve it. Ultimately, however, the aim of proposal writers is to persuade the reader to place the proposal on the short list, and finally to choose it as the winning proposal out of all the bids that are submitted.

8.3.2 Eight key discourse topics

Figure 8.2 depicts the macro-discourse functions performed by engineering proposals generally. It also shows eight main discourse topics,

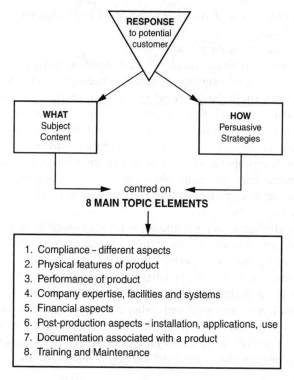

Figure 8.2 Macro-level discourse functions and topics of technical proposals

referred to as topic elements in this discussion, that relate to proposal writing. From the engineers' standpoint, proposals need to persuade the customer about the following eight key topics:

1. *Compliance*

This concerns different aspects of compliance, for example, degree of compliance. The proposal needs to show how closely the proposed 'solution' matches the customer's requirement (see Chapter 1 for a discussion of 'solution' vis à vis 'product'). Ideally, it should meet all the criteria set, or it needs to convince the customer that the product would be best for the customer's purposes, even if it does not quite fit the bill (or in engineer-speak is not 100% compliant). So it needs to explain convincingly what is being proposed, and why. Compliance is a crucial topic because it may be the most important selection criterion and overriding factor in the customer's selection process.

2. *Physical features of the product*

This topic element focuses on physical construction and appearance. If a piece of hardware is being proposed, the proposal provides a description of the equipment, in particular, what it is made of, how it is made, and its appearance, through text, supported by photographs and diagrams galore. If a piece of software is being described, in other words, a kind of 'abstract' product, with textual (or conceptual) rather than physical substance, the proposal includes details of the casing/cabinet/console that contains the electronic data (sometimes nothing more than a mundane plastic cartridge) with pictures of that instead!

3. *Performance of the product*

This topic element conveys information often as scientific or mathematical proofs, and is one of the most difficult (and riskiest) topic elements to include in proposals, especially if the product does not exist. When it does exist, engineers describe what the equipment is capable of doing and how it has performed in tests, and may also provide information about how it has performed with other customers who have bought it. The better it performs (for the price), the more likely the customer is to buy it. However, if engineers promise more in the proposal than it is capable of achieving, they must develop it further, at the company's expense.

4. *Company expertise, facilities, and systems*

Since the customer may not know much about the company submitting the proposal, he needs to be persuaded about its suitability for being awarded the contract. Whether he lives outside or within the company's own country, he may need to be given information about company structures, personnel, manufacturing facilities, testing facilities, and so on. He needs the company's measure, if he is to be convinced of the value of the solution being proposed. For big contracts, the customer also needs to be reassured of the company's financial stability and robustness (at times this information is gathered covertly). It is not uncommon for disputes to arise during such contracts, and the company needs to be financially healthy enough to withstand them, which may prove costly indeed.

5. *Financial aspects (price)*

Giving a price breakdown is a crucial function of any proposal. The calculations and presentation of financial aspects are done by colleagues with commercial responsibilities (who may have been engineers in a previous working life). A proportion of technical proposals make a reference to price, sometimes implicitly, but much depends, of course, on whether the proposal is an all-in-one volume, or split up into separate sections. Engineers provide estimates of costs, working out what they believe will be needed to develop the product. They pass these estimates on to the accountants, who 'work them up' into a price. This price is then passed on to the commercial department, where commercial colleagues work with managers to draw up a 'market price'. It happens sometimes that this 'market price' is less than the engineers originally estimated, which is 'when the trouble starts', according to one engineer, who added: '... but the engineers aren't good at winning arguments with sales guys!'

The commercial and financial sections deal with the financial implications to the customer of particular phases of the project, in the event of the proposal winning in the bid process. Those in the bid team in charge of the finance will want to ensure the release of staged payments to the company when certain 'milestones' are reached. These are particular stages detailed in the proposal, which may concern the achievement of objectives in product development, particular tests being carried out, the successful production of a proto type, and so on. These sorts of financial consideration underlie the processes of design, so that aspects of proposal discourse reveal evidence of what can be referred to, for want of a better term, as monetary manipulation. This is manifest, first and less obviously, in the organisation of the discourse, and secondly, and overtly, in direct reference to financial aspects in the proposal. The following are post-production aspects:

6. *Use of the product*

This is relevant in the post-production phase and is where the user, as distinct from the customer, becomes the focus of attention. The user's perspective needs to be reflected in the proposal, which may need to show how easy it is to install the product, how easy it is to learn, how many people are needed for it to function, how safe it is, and how easy it is to maintain. In-service training courses for staff may be part of the proposal. Also, it needs to be known what spare parts are available, and how often it will need checking and maintenance. The next two elements are similarly user-orientated.

7. *Documentation*

This element relates to all the documentation associated with the product, sometimes referred to as user-documentation, documentation support, or, in the case of electronically stored information, online help.

The proposal may need to describe the support to be provided in the form of servicing, staff training manuals, and user manuals, which may be paper-based, on compact disk, or online.

8. *Training and maintenance support*

The bid may need to include information about engineer advisors and trainers, maintenance and trouble-shooting engineers who will be on hand in case of any emergencies, equipment spares, and service facilities. Sometimes, potential for upgrades is mentioned.

These, then, are the eight main topic elements that are usually included in technical proposals, the first four being obligatory, and the last four optional, depending on the nature of the solution being proposed. To gain further insight into proposal structure, it is useful at this point to remind ourselves that these topics are being examined from a persuasion perspective, since the ultimate function of proposals is to persuade. Put simply, engineers attempt to persuade by employing particular persuasive strategies, or 'suasion', to use Bolinger's term (1980: 111). Discussion in Chapter 6 shows how persuasion is multi-faceted and complex, but in the engineering context it is possible to tease out three purposes for the strategy (or suasion) employed. It can be seen that by their nature they are not mutually exclusive, and so there is a degree of overlap between them. The three purposes are these:

1. *assurance and reassurance* – to make the customer and other readers feel secure about the solution, and to rid them of any concerns about a variety of topic elements, ranging from product description to quality assurance procedures.
2. *'face building' and corporate-image building* – to impress readers about the engineering team's knowledge and expertise, and about the company's facilities and connections.
3. *instruction* – to impart knowledge and information to enable readers to understand the product and its esoteric features, and sometimes to educate readers about technical/ scientific matters.

These underpin the strategies for persuading the customer in proposals, influencing engineers' decisions about textual matters, that is, what should they write (or say) about the solution, and how should they present it. Figure 8.3 is an attempt to provide a persuasion 'map' for the sections of proposals that concern the engineers. It provides more specific information at a glance about the discourse strategies that are used by engineers, the topic elements that lend support to them, and the

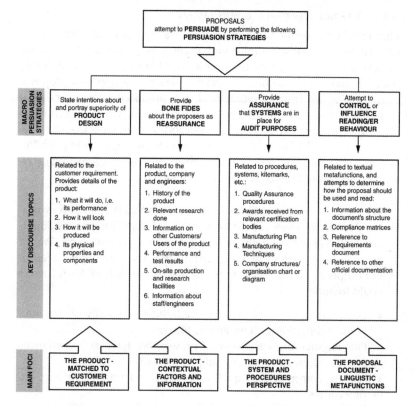

Figure 8.3 Persuasion strategies in engineering proposals

ways all these are linked to particular aspects of the engineering product. The first four of the eight topic elements listed above are incorporated to show how they are manifest linguistically in proposal text to assist engineers in their attempts to influence the behaviour of the customer, through assurance and/or reassurance, instruction, and 'face-' or image-building.

8.4 Themes

Themes merit more attention than they have received in the literature. The little that has been written about them is now examined, as it provides useful insights into the kind of complex decision-making that goes on when engineers prepare proposals.

8.4.1 Themes according to Ellis (1997)

Ellis (1997) is one of the few in the area of technical communication to provide information about themes in proposals. Like Stross, discussed below, he mentions the role of 'themes', a term commonly used throughout the industry to refer to selling features of a proposal. Ellis's section on themes, albeit brief, is reproduced below, because it sums up rather well a particular view on themes, which is common-sensical though rather general in applicability. Engineers would recognise some pet phrases in the extract, which are commonly found in engineering proposals, for example 'ensure low risk', 'engineering excellence', 'committed to quality', and 'track record':

> We want our readers to be aware of certain broad themes as they read our document. These are the keys that will help to unlock any doubts and establish our credibility as to why we should be selected. As they are key themes we must make certain that they are appropriate for our purpose and that they are repeated with conviction. Such themes could include:
>
> - Our approach is evolutionary; we build successfully on previous work and by doing so we ensure low risk.
> - Our engineering excellence is proven; we have an experienced systems team enhanced by subject specialists.
> - We are committed to quality. You are welcome to inspect our procedures.
> - We are a small flexible operation and can react with speed to situations; our track record demonstrates this ability.
> - We consistently meet deadlines. (Ellis 1997: 169)

8.4.2 Themes according to Stross (1990)

In a similar vein to Ellis, Stross (1990) discusses themes as an essential and early part of proposal writing, also structuring his advice as bulleted points. Significant numbers of engineers have attended his training sessions on proposal writing and found his advice useful, since he attempts to address a fundamental problem engineers have in proposal writing that concerns the information content of proposals. Like any writer with the task of producing an extended exposition, selection of topics presents a major problem: engineers need serious help in deciding what to write about. It is for this reason that a rather fuller account of Stross's work in this area is included than might be expected (Figure 8.4),

Garden-variety themes:

- The total resources of the firm will be brought to bear on this programme to ensure its successful completion.
- Our past performance is the credible base for the proposed performance.
- We are high performance, low risk.
- We have exclusively served the Navy for 30 years.
- We built it before; we can do it again.
- Our proposal is 100% responsive to the government's requirements.
- We are low cost, low risk with production data available to prove it.
- Our state-of-the-art leadership assures lowest technical and cost risk.
- We were the contractor on your last programme which served as a basis for this one.
- To us, this is not just another contract, it is the only one we have served on for 37 years.
- In total, our firm has over 10,000 person-years of experience in this field.

Configuration themes:

- This aircraft design will fully serve the common mission needs of both the Air Force and the Navy.
- 44% of our proposed design in common with prior product.
- Commonality with A-6 engine – engine now in production.
- Reliable systems archived through design with back-up and quality assurance.
- Our design features lowest acquisition cost and lowest maintenance cost.
- Our proposed approach will reduce the number of government personnel required overseas.
- We have verified our design through extensive testing.
- Our design has proven long life in operational use.
- We will modify available government-furnished equipment from previous contract.
- We have a proven and fully-tested design approach.
- We are so sure of equipment reliability that we'll assume the risk of maintenance and a fixed price.
- Alternative design approaches have been identified if problems arise.
- Trade-off studies demonstrate superiority of proposed design.
- We feature low development risk in our approach.
- Fewer components.
- Use off-the-shelf components.
- Our approach offers low cost, high reliability.
- Unique design.

Competitive themes:

- As the incumbent contractor, we have our first team on the project now; there will be no disruption of morale or performance.
- We have world-wide capability to service this project.
- We have unique in-house IR&D directly related to the technical solution.
- Only firm with design experience under nuclear environment.
- We are the only minority-owned firm capable of doing this job on time, within budget.

Figure 8.4   A selection of Stross's themes (Stross 1990)

- Our local office will facilitate coordination of the project and promote communication between contractor and government personnel.
- Our project team features a single-line chain-of-command straight from the government technical
- representative through to our field personnel.
- We would draw upon our existing manpower permitting a rapid programme start.
- We have no overrun history.
- We have more experience in working with your agency than any other single firm.
- Two years of IR&D directly applicable to 95% of this requirement.
- Our team represents every Congressional district.
- No additional facilities will be required for this programme.
- Leading experts in the nation.

[Here are examples of Stross's throw-away themes, which he believes will not achieve success for a proposal, if they are used solely:]

Throw-away themes:

- Our proposed project director has full authority to command the full resources of the firm for this contract.
- We have top management's backing for the project.
- We will use our existing management team.

Figure 8.4   (Continued)

although it is but a short extract of his material, taken from a course handbook compiled for training purposes. The extract should be read with this in mind because, as they stand, the themes lend themselves to further refinement.

The value of Stross's work is his demonstration of an attempt to devise a comprehensive list of themes and to categorise them. A few of the themes are expressed in words that some associate with an unabashed kind of 'sales-pitch' but, nevertheless, Stross's lists are well liked by engineers. They find this mode of presentation attractive, because they are provided with writing ideas, or topics, in a form akin to a check list or menu that they can pick and choose from.

Stross sees four categories of theme, three major and one minor, which are as follows:

Major theme categories
1. 'common garden variety' themes
2. themes 'unique to the bidder's configuration' (sometimes called 'discriminators', 'configuration' being used differently from the way engineers would use the word)
3. 'competitive' themes

Minor category
4. 'lesser throw-away' themes.

Although the categories may not withstand rigorous examination, they nonetheless represent a creditable attempt to identify a notion that may be referred to as the 'what' factor or the 'aboutness' of text (Marder 1960: 61). 'Aboutness' concerns the information content of text, and in the case of proposals, relates to what engineers need to write about in order to persuade a potential customer to choose their solution.

However, some engineers would take issue with the examples used and language adopted by Ellis and Stross, because they see themes as super-ordinate, over-arching selling points, functioning in proposals much like mission statements in management practice, or learning aims as foci for learning objectives in education institutions. One of them explained a theme thus, as an illustration, reproduced verbatim ('end user' is underlined, as it was said with particular emphasis):

> The Seahunter mid-life update might be an example [of a proposal]. We proposed a very expensive LongueMadison thermal imager as central to our solution. The theme was 'only the best is good enough'. In the trade off between cost and performance, we went for performance and meeting the stringent specification fully. The system is an enhancement to protection against fast, sea-skimming, anti-ship missiles. If the system doesn't do the job – you lose your ship.
>
> The customer didn't go for it – he would rather lose the ship, but save a few bob! Of course the customer isn't sitting on the ship – the <u>end user</u> is!

It is possible to see links between two of Stross's categories, with the major categories of Proposal Components (PCs) discussed in this chapter: most of his 'garden-variety' themes relate to personnel- and company-related topics; his configuration themes more or less relate to product or design matters; items in his 'competitive' category seem to be a mixture of the previous two. However, the rationale for 'throw-away' themes is difficult to determine; perhaps they should be re-allocated, or, true to their type, be discarded. In contrast with the themes identified by Stross, Ellis's list relates only to company- or personnel-related features of the proposal, although no significance can be attributed to this, considering the brevity of the section on proposals in Ellis's book.

8.5   The identification of proposal components (PCs)

8.5.1   The search for an analytical framework

Certain kinds of text analysis, especially those concerned with communicative purpose, have been well served by a genre approach to text analysis. For teaching and learning purposes, particularly in the areas of English for Academic Purposes, English for Business Purposes, and English for Science and Technology, pioneering work has been done in the area of academic writing, particularly genre analysis (Swales 1981). Teachers and students at universities have found a genre approach to text to be illuminating and useful in their own academic writing. It is especially effective when applied to shorter texts that are written strategically, usually by individuals (Swales 1990, Bhatia 1993). However, this approach is difficult to apply to longer texts, like the discussion sections of MSc dissertations (Dudley-Evans 1986: 144), and has proved less effective with the longer proposal documents produced by engineers. It has been necessary to look beyond well-trodden paths in the genre analysis domain, to provide descriptions of interest to engineers and useful in their work, without requiring a pedagogic interface.

The discussion that follows has as an underlying theme which is reminiscent of Sinclair's ideas about 'open choice principle' texts. These are based on a 'slot and fill' principle, which Sincliar regrads as a segmental view that is more relevant to specialised texts like legal or technical texts (Sinclair 1987: 320, 324). In their work on texts in scientific textbooks, Davies and Greene seem to take a 'slot-and-fill' view, as revealed in this explanation of their analytical approach:

> Each example has been analysed to show how the text 'fills' the slots of the information structure or frame. (Davies and Greene 1984: 130)

The idea that text can be constructed or deconstructed in much the same way as a machine can be assembled or taken apart is appealing to engineers.

A 'master' genre, matrix, or colony

There are parallels to be drawn between writing in academe and the engineering workplace. Swales and Feak (1994: 157) describe the overall rhetorical shape of the research paper as comprising four different sections, each section having a different purpose and distinctive linguistic structure: Introduction, Methods and materials, Results, and

Discussion. The labels or headings for these sections that they use would be easily recognised by any graduate student.

The engineering proposal is similar to the research paper in that it comprises a collection of sections, each of which is distinctive in terms of purpose and structure, and each of which contributes to the construction of the whole proposal. In fact, it could be seen to be a kind of 'master' genre, or master document, since it is a textual compilation of sections that, in turn, comprise other sections and segments. The concept of a 'matrix' is also attractive here, since a matrix (or rather, a textual matrix) is a holding structure that provides a textual environment for an array of texts that make up the proposal. Although matrices are usually associated with the field of mathematics, it can be seen that the idea of a matrix is also applicable to documents which describe engineers' solution to the customer.

Any thoughts about matrices in this book have been inspired by Hoey, who developed ideas throughout the 1980s about text colonies, together with discussion of hierarchical organisations of texts in his book on written discourse analysis (Hoey 2001). Both Hoey's ideas about matrices and text colonies are apt for several reasons, one being that the proposal may be accessed at different points by different readers, who may read only those parts that are relevant to their reading roles. Putting aside for a moment the trouble engineers have with writing persuasively and selecting topics to write about, proposals present a special problem simply by dint of being such big texts.

Dudley-Evans, who was interested in the analysis of longer academic texts, observes that 'one of the greatest problems' is 'the very long informing sections that so often occur in the middle of articles, dissertations and lectures' (1986: 120). Like Dudley-Evans, Hoey has an appreciation of the difficulties of writers 'losing their sense of the overall picture' in the case of longer texts (2001: 52). The model he proposes is a modification of Pike's use of a matrix to represent the structure of an event, or 'happening', as Hoey puts it (Pike 1981, cited in Hoey 2001: 93). His analyses are mainly concerned with narratives, but his portrayal of the matrix perspective seems apt indeed for this study. His concern about the route followed in the 'telling' of a narrative, whether it passes across or down the rows of a matrix, is less central to this discussion than the idea of a document comprising segments (or 'cells'). This has proved the most useful view to take, since engineering proposals display structural features similar to those observed by Segal, whose analysis of a medical review article shows:

Parts are not organized with persuasive introduction and discussion sections flanking a more descriptive middle; rather, they are arranged according to topics, with headings and sub-headings directing reader attention to particular areas of professional interest. Review articles, however, are not less persuasive because of this seemingly arhetorical organization. (Segal 1993: 94)

It is from a blend of Hoey's matrix and colony perspective that the PCs described in this chapter have been developed.

8.5.2   Towards a more topic-focused description

Since this study aims to be of some use to engineers, it was decided to tease particular persuasion topics which may have the potential to be developed by engineers, or applied linguists who have an interest in such texts. A concern shared by both groups lies in the selection of information to be used as 'ammunition' for convincing the customer of the efficacy of the proposed solution. What should be included in the proposal? The 'what' factor is fundamental to any drafting of text (Hopkins and Dudley-Evans 1988: 113) and, in this respect, rhetorical discourse in the engineering workplace is little different from that in other domains. Deciding what the text should be about, the 'about-ness' of text, mentioned above, is crucial. Invariably, one of the highest ranked problems for writers of this kind of rhetorical discourse is the gathering and selection of information to be included.

The strategy of providing 'hard information' in text in order to persuade is particularly important in engineering proposals. However, it demands an approach to analysing texts that has tended to be neglected by applied linguists, possibly because it requires understanding of the discourse community. It has already been explained in earlier chapters that engineers are not comfortable with overt persuasion. They would certainly find it distasteful if there was even a hint that they may be regarded as boastful, and so they try to impress in less obvious ways, for example, through listing 'facts' about their work. The following extract from a proposal shows instances of engineers trying to impress without being obvious about it. Chapter 6 discusses studies of the restrained kind of persuasion practised by engineers. Information gathered from this work helped in the identification of particular topics that are useful for engineers to include in proposals. The idea behind this work was inspired by the lists drawn up by Stross (1990) and observations of the enthusiastic response they received from engineers. The analytical process that this involved was complicated and drawn out, based on a close analysis of the information content of 11 engineering proposals and accompanying executive summaries. A corpus of 95 proposals was used to check findings and gain further insights. The analysis involved, among other methods, drawing inferences from proposal texts with the help of specialist informants (the engineers). Let us consider, as an illustration of a small part of the process, the following extract, which is intended to impress the customer about a company's position in terms of development, production, and sales in the gyroscope field. The number of each sentence is shown at the end of the sentence in brackets.

A LEADING EUROPEAN SUPPLIER OF INERTIAL PRODUCTS
SPACETRONICS has been a major supplier of inertial sensing products and systems for over 80 years and is the foremost manufacturer of such products in Europe (1). Current gyro throughput is in excess of 500 per month (2). This proposal for the use of a Vibrating Structure Gyro (VSG) is based on over 10 years of VSG development and over 5,000 VSG sales (3). The new technology Silicon VSG is an evolutionary step in the VSG development progression and demonstrates a high level of innovation, setting the benchmark for others to follow (4).

Although a brief extract, particular 'impressive' features can be inferred, for example:

1. company's standing (Whole paragraph)
2. good reputation of the company (Whole paragraph)
3. impressive size or scale of development and production (sentences 1–3)
4. state-of-the-art features (sentence 4)
5. expertise and experience of personnel (sentences 3–4).

Discussion with an engineer revealed another slant to the inference concerning the 'track record' or reputation of the company and gave rise to the word 'pedigree', which proved useful as a name for one of the PCs. He drew an odd analogy to illustrate his point:

> Well, they're talking about our pedigree, aren't they?...You could be running a stud farm and end up with a horse that's really ropey, but you could go back and trace its lineage and that's its pedigree.

The term 'pedigree' is often used by engineers, although it usually carries more positive connotations than that of a 'ropey horse', to convey a picture of a committed, experienced, and expert engineering company. Discussions like this were useful in guiding my attempts to devise a taxonomy of analytical categories to better understand the structure of proposal documents. This particular example indicated the existence of an information 'topic' relating to a company's (or engineers') reputation, and these topics were developed into PCs, as they later came to be called.

Out of such features, a comprehensive list of 39 individual text-information segments was identified, each segment reflected by labels that expressed distinct, discrete topics. These segments, called PCs, are listed under four main discourse function categories:

1. *Product or solution-focused PCs.* These concern aspects of the product (or solution, as the engineers sometimes refer to it) specifically, for example its design or performance.
2. *Company-focused PCs.* These convey information about the company submitting the bid, its collaborating partners (other companies), and its personnel, which in the case of proposals is most often the engineers concerned with the design and after-sales support of the product.
3. *Product and customer-support-focused PCs.* PCs for this category cover any aspect of the service offered to the customer after he has taken delivery of the product. Product support is otherwise referred to as ILS, customer support, or after-sales support. This is significant in most projects, especially those which may last a few years, or even extend over one or two decades, although this aspect tends to be under-represented in proposal text.
4. *Metalinguistic-focused PCs.* This is a group which is markedly different in nature from the three above because it is text-oriented rather than topic-oriented. It relates to discourse and text organisation, and assists in influencing how the document is read. These PCs are used to indicate how the whole proposal should be read in terms of both reading behaviour and reader attitude. Metalinguistic PCs help the writer to refer to the document itself, parts of the document, or to relate these document parts to other documents. Since this function may relate to substantial stretches of written discourse, it is also referred to as 'meta-discoursal'.

The PCs in Figure 8.5 could potentially be found in any engineering proposal, although this does not usually happen. Logically, engineers need to decide on the central theme(s) for a proposal, before deciding

1 **Product (or Solution) Focus**
   Compliance - how compliant or degree of •
   Cost benefit or implication, including potential savings •
   Costings
   Design improvements or special features •
   Low risk •
   Manufacturing plan
   Off-the-shelf aspects •
   Packaging
   Potential improvements or benefits to the design •
   Product or solution gloss (or statement) •
   Programme schedule
   Proven performance and tests success •
   Risk
   State-of-the-art features •
   Technical response to requirement specifications (may be in Appendix)
   Testing and tests
   Track record of the product •
   Viability/Feasibility/attainability of the solution •

2 **Company Focus, including engineering personnel**
   Collaboration and alliance benefits •
   Company pedigree and/or reputation •
   Company's commitment •
   Company's good **R** and **D** facilities
   Company structure/information/profile
   Expertise and experience of personnel •
   Kitemarks, standards, and accreditation
   Quality Assurance
   Production and manufacturing - good facilities and high standards •

3 **Customer (and product) Support Focus**
   Aftersales customer and product support provided •
   Availability Reliability Maintainability
   Integrated Logistics Support (ILS)
   Installation
   Maintenance and support in the field •
   Training provision
   User documentation, manuals, online help, etc.

4 **Meta-discoursal Focus**
   Distribution list
   Indication of particular docsections •
   Proposal title
   Referential or context setting •
   Security rating of document
   Copyright

Key: • Indicates PC also appears in executive summaries

Figure 8.5 Taxonomy of proposal components (PCs) under four main 'what' categories

which PCs most effectively reflect the key 'selling points' of the solution. All are clear topics that may be presented as 'benefits' to the customer, with the exception of the final category of metadiscoursal PCs. Since the nature of the solution will determine the most suitable PCs for inclusion in a particular proposal, it follows that any proposal will contain a selection.

8.6   Conclusion: reverse-engineered text

This approach taken is a reversal, more or less, of the process engineers follow when compiling the documents, although the analytical process proved rather more sedate and systematic than the writing of them. Also, the analysis was carried out over 5 years or more, whereas engineers have to produce proposals within very short time-frames. Chapters 4 and 6 show the constraints engineers are placed under, when they have little time to produce documents like proposals. They may have only two weeks to produce these large texts, or a month at the very most, usually. Therefore, proposal writing can be frenetic, with engineers wishing they could be more in control of the whole process, and more informed when selecting and structuring information to be included.

The motive behind doing an analysis of proposals has always been to produce a description that engineers would find useful in some way. The approach taken was a conscious attempt to cater to their interests and needs. It was clear that the description should convey the essentially persuasive intent underlying proposals, as well as the 'aboutness' of the texts. A grounded approach was followed from the start (see Chapter 1 about this), which enabled information to be derived from a bank of proposal texts and from the engineers who wrote them. The analytical process was therefore 'bottom-up' in that the texts were analysed with few preconceptions on my part, and it was the texts themselves, and the engineers, that eventually yielded the PC categories.

The 'reverse engineering' of texts has meant, in effect, that proposals have undergone a deconstruction, with the focus being on information content and writing motive(s). This process has yielded two important kinds of output as a result:

1. Information about the substance of proposals, that is, their topic content or the 'what' aspect. The analysis has shown that the information in these documents can be accounted for by a closed set of 39 information components, called PCs, belonging to four mutually exclusive information categories.

2. Definitive lists of PCs that combine to form text for proposals and executive summaries. Data derived from PCs and their realisations can be used to produce graphical representations of the texts, in an attempt to recognise patterns which would otherwise remain undetected. These patterns and their implications are discussed in Chapter 9.

3. Realisations for each PC in the form of text segments, comprising a word, a phrase, a paragraph, or longer stretches of text. PCs are useful for identifying particular text segments to enable coding (or tagging) to be carried out for further analysis. There is not the scope in this book to provide a proper description of such treatment of text, although there is some discussion of this in the next chapter. Aspects of an analysis of proposals and executive summaries using PCs as a basis are discussed in Chapter 9.

9
Executive Summaries

9.1 Introduction: what are executive summaries?

Essentially, the executive summary is a persuasive abstract that accompanies a proposal when it is submitted to the customer. Engineers usually submit an executive summary with most proposals, regardless of their length, even if the customer has not asked for one, since it is considered courteous and a good strategy to provide one, for reasons that are discussed later. A good executive summary encapsulates the whole proposal, and therefore ideally serves as a text through which the whole proposal can be considered in microcosm. Writers aim to make it a convenient length: it is usually a short text of around one or two pages, although those for large proposals may be significantly longer, reaching 16 pages and around 3000 words in length, or more.

9.1.1 Purpose of the executive summary

Engineers tend to suspect that the executive summary is a more important element of the proposal than is generally credited. The summary's main purpose is 'to highlight the key benefits of choosing our company's solution', to quote an engineer's words, the benefits being, in essence, information topics or main selling points, around which a proposal is constructed. Usually these topics relate to features of the product that the engineers believe would be attractive to the customer, or advantageous for him to have.

An executive summary should serve as a good representation of the proposal, summarising its key selling features, or 'benefits', which is a term engineers prefer to use when discussing information from the customer's standpoint. It may be pivotal in the bid process, by persuading the chief executive for the customer of the efficacy of the proposal.

Linguists talk about persuasive language and persuasive texts, although it would be more precise to say that executive summaries are written with persuasive *intent*, rather than that they are persuasive. Strictly speaking, no text is persuasive unless it has succeeded in changing readers' behaviour or attitude in the way intended by the writer. Regrettably for the writers, however, only a fraction of executive summaries are ever successful, simply because of the nature of the bid process. Proposals and notions of persuasion are examined in Chapters 6–8.

A courtesy convention to ensure polite bidding behaviour

There are gaps in engineers' knowledge about the value of the executive summary and this has made it difficult for them to identify its function. This situation is not helped by the fact that, at times, it seems to be an optional element and, at others, obligatory, with the customer sometimes specifying that one should be submitted and, at other times, not. It is generally accepted that a summary should be submitted as a matter of courtesy or demonstration of politeness to the chief executive, who, by dint of his position, would be the official recipient of the proposal documents. If a summary is not submitted, it is believed an equivalent textual gesture, in the form of a cover letter, for example, should be submitted as a matter of polite bidding behaviour. As previous chapters explain, proposals tend to generate anxiety amongst engineers, who try to make sense of the anecdotes and gossip a secretive event like this inevitably generates. They talk together and surmise about post-proposal-writing scenarios taking place at the customer's location. Although specific knowledge about the purpose of the executive summary is elusive, the upshot of these discussions is that it is a key document. This acceptance that it is an important part of the proposal is based on the following reasons:

1. There is a traditional regard for them as texts with status, albeit undefined, because of their association with chief executives;
2. It is important to relay to decision-maker(s) and/or the chief executive the proposal's main selling points in summary form.

Rumour and mystery in a secretive climate

No one really knows what happens behind closed doors when bids are being selected during a tendering process, unless he or she happens to be a member of the inner circle, representing the customer and scrutinising the proposals that have been submitted. In the final stages of writing a proposal, the limited time that usually remains for writing the executive summary gives a false indication of its importance, for, and engineers readily acknowledge this, the summary rarely receives the time it deserves. When reflecting on their writing practice, engineers have said that, in the pressure to produce the proposal documents in time, the executive summary may be left until last, being squeezed in and hurriedly written in the final stages of proposal preparation. More often than not, this is the case.

It has been suggested that the chief executive's minions consider the submissions on his behalf, and that he usually follows their recommendations, in which case the executive summary would not be as important as the main documents. However, this would be rejected by those with knowledge of the bid process from the customer's perspective: it is not unusual for engineers to have worked for other engineering organisations or to have been ex-service personnel, with some (at least anecdotal) knowledge about procurement. Also, companies themselves invite proposals from other companies, and will have procedures for examining and short-listing them.

Rumour and mystery surround the executive summary, simply because the whole bid process is shrouded in secrecy. That it has to be so, to maintain the integrity of the process, means that engineers engage in guesswork about the extent to which the executive summary contributes to the whole proposal. They surmise that, by virtue of its name, it might be read by the chief executive, whom they believe to be one of the key decision-makers in selecting the winning bid. However, there is uncertainty about how the proposal is read, or how many people are involved in the selection process for weeding out submissions to form a shortlist. It is usual for several readers to be involved in large projects for which lengthy proposals, comprising several volumes and addenda, are written. (The bid process is discussed in more detail in Chapter 6.) However, there is less knowledge about executive summaries. Engineers generally have only hazy ideas about how wide the readership of the summary is, or the purpose it serves.

'Transparency' and 'openness' there still is not

Overall, then, engineers would agree that the executive summary is a textual thorn in their sides. They are persistently concerned about it, and proposal writing generally, and this causes anxiety in some cases which exacerbates what is an already stressful bid process. This concern is doubtless rooted in the fact that they can only guess about what happens to the proposal after it is delivered to the customer. The few textbooks dealing with proposal writing and giving advice on how to approach it (e.g., Ellis 1997, Newman 2003) make clear the complexity of the process from the proposal writers' viewpoint. There is a general recognition of the need for transparency and openness, and indeed the publication of detailed selection criteria, processes, and procedures by those putting contracts out to tender gives the impression that there is clarity of process. However, this is a false impression, since little is known about what happens once proposals have been submitted.

It is possible that little has changed since Marshall (1986) describes customer behaviours in the construction industry, which showed unofficial (and, some would say, illicit) price bargaining to be an integral part of proposal short-listing practices. Who knows and can say? Even today, with rigorous bid procedures in place, engineers remain in ignorance about much of what happens between the time they deliver the proposal to the customer and the moment the winner is announced. Judging by the compartmentalisation of working practices and secrecy surrounding proposals, not only in engineering but across other sectors as well, engineers' scanty knowledge of proposal reading practices, and ensuing deliberations by the customer, is therefore understandable.

Research journal entry: Questions about ethics in the post-'cost-plus' bid process

More about proposal vetting. I talked about my research in Southeast Asia just now, based on documents in the building industry, and how the tender report summary was used by the customer to beat down prices during tender interviews. Apparently, this goes on here as well, although the engineers don't seem very comfortable with it. I would have thought this was rather dodgy practice but it is openly acknowledged. I see Newman writes about pre-ordained [or fixed] decision making, referring to selection that is 'wired' (2001–2003: 10). I recall how the quantity surveyors tried to be honest brokers, but used to regard [wearily] as inevitable the claims, counter claims and court cases that were an inevitable part of every huge project. They used to find them exhausting.

When I told Frank this, he said this sort of bargaining goes on in this country too. Once they submit a proposal, the customer

---

Continued

may contact them and try to beat them down through a series of questions or phone calls [eyebrow raising, I would have thought]. Although the MoD is more professional in its practice, it seems. In the old 'cost-plus' days, it didn't matter what the eventual agreed price was so far as MoD (and DoD) contracts were concerned. The contractor got paid his costs plus 10%. That all ended in the late 1980s in the UK, although it is still the case with DoD contracts and US companies today.

Frank mentioned one project where the winning company was able to reduce its price from £30 to £20 million. He reckoned there would be real problems further down the road, though, and was in no doubt they simply couldn't deliver at that price. Short-term strategy for long-term [not gain but] losses for both parties!

So things haven't changed much. Any dodgy practice rather makes a mockery of all the hard work that goes into proposal writing, and must dishearten engineers who try to be creative and professional in the solutions they come up with. Must lead to cynicism surely?

9.2 Structure, information content, and presentation of executive summaries

9.2.1 The summary as a notion

Summarising is an everyday skill we usually take for granted, most commonly exercised in everyday chit-chat. When we summarise a film for a friend, say, we are selecting the most important bits of information that, in our view, we feel the friend should know, or would like to know, depending on our own judgement of the situation. It may be that the friend would find the summary useful in order to decide whether or not to see the film. However, some novice raconteurs, unaware that the best film summaries are brief, provide instead drawn-out blow-by-blow accounts of a film's narrative. Anyone who has suffered such accounts will know these can make tedious listening.

A high-order skill

A summary usually presents the substance of a text in a condensed form. They comprise a selection of the most salient or pertinent pieces of information (or points) contained in the original, although the decision about which information should be chosen rests with the summary producers, who have to exercise their judgement about such matters. This is why summaries are considered difficult: it is no easy task to identify these points or to present them as a condensed version of the original. This is another quality of a summary: it is intended to impart the essence of the story (or document) more quickly, and so it usually follows that it should be shorter than the original. Brevity, direct relevance to, and a true representation of the original are key qualities of a summary. If the summary enables the recipient to better make a decision about a matter, it will have achieved its ultimate communicative purpose.

The ability to write an effective summary requires a high order of skill and thought. That it is intellectually demanding has long been recognised by educators, who have seen the value of summary practice in teaching for developing rhetorical skills in students (Swales and Feak 1994: 105, Grabe and Kaplan 1996: 325). Summaries (and précis) have also long been included as examination questions, blighting the lives of countless generations of school children. Engineers summarise in informal situations as well as anybody else; however, they find it a different and difficult matter to produce formal summaries in written communications. The executive summary endures as one of the most difficult documents that engineers have to write at work.

9.2.2 A neglected genre

Executive summaries have attracted little attention up to now, although information about them can be found in guides for proposal writers (Covey 1997: 405, Newman 2003: 44). Covey, for example, provides advice to those (non-linguists) in commerce and industry about producing a variety of business genres, and suggests a range of layouts and styles for the executive summary. Both publications are popular with engineers and technical authors, who like the prescriptive character of the books, and consider them to be good reference material of the genre. Covey provides four exemplars (model texts) with commonsensical side-annotations indicating key aspects of the summary, for example the need to keep paragraphs 'short and focused', to end every sub-heading with a section number which should cross-reference to corresponding sections in the proposal, and so on. He uses the terms 'benefit' and 'theme', mentioning that the summary should list 'three major benefits covered in the proposal'. It is clear he assumes the reader knows about these, since they are not explained. However, engineers are not always clear about themes and benefits, and find it no simple task to make a selection for inclusion in the executive summary. This dearth of information about executive summaries is surprising, considering their influential role in the selection process and the amount of effort (and cost) that is invested in proposal writing generally. It is probable that the secrecy surrounding the inner-workings of the tendering process, mentioned earlier, and the restricted access that has so far been afforded to researchers, have contributed to the neglect of this genre. Considering how little has been written about executive summaries, then, this chapter provides some idea about how these summaries may appear and different modes of presentation. Their different discourse functions are also identified and described with a view to providing insights into this text-type that engineers find troublesome.

9.2.3 Persuasion through selling points: discourse functions

An examination of executive summaries has revealed the existence of distinct selling points, which may be regarded as generic (Sales 2002). In theory, an executive summary could mention all the selling points featured in a proposal. Since proposals differ, however, and every competitive proposal is distinctive (usually unique), the actual selling points mentioned varies from bid to bid. An executive summary may mention several selling points, or only one or two, as will be shown, and they usually appear in order of importance, as ranked by the bid team, the most important mentioned first.

Chapter 8 provides a fuller discussion of selling points, showing that they convey particular information about the solution being proposed to the customer, and giving a rationale for referring to them as PCs. Figure 8.5 provides a complete list of PCs, showing all the PCs to be found in proposals and executive summaries. As Chapter 8 explains, although they are not intrinsically selling points, metadiscoursal PCs are also included in the list of PCs. Having identified the PCs to be found in executive summaries, we can now consider how they are organised and realised in authentic texts. First, it may be useful to consider a typical (or generic) structure of an executive summary, before looking at actual examples. The generic structure is portrayed twice: as a pared-down outline diagram (Figure 9.1) and as a table (Figure 9.2) containing actual examples of text components, or PCs as the components will be referred to from now on. The former shows how PCs are organised as distinct segments of executive summary text, and Figure 9.2 provides realisations of PCs in different functional categories, as authentic examples. Figure 9.1 combines findings from the data

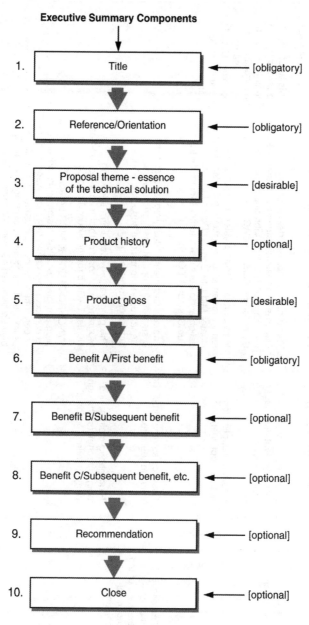

Figure 9.1 Outline generic structure of the executive summary

Functional component	Discourse function	Example extracts taken from executive summaries in the corpus	Comment
1. Title	Labelling (Obligatory)	**Executive Summary / Executive Brief**	Usually appears as a main heading.
2. Reference/ orientation	Reference provision – locates/orientates reader (obligatory)	This proposal provides details of the Osborne-Marshall Industries (OMI) and Leedor Products (LPP) Silicon Vibrating Structure Gyro, for the Barmstedt AG.	This would be regarded by engineers as an introduction, or some kind of opening. Usually a sentence-long paragraph.
	Courtesy/supplication marker (optional)	In response to MME Systems invitation to tender for an Electro-Optic Sub System for the Hunter Mid Life Update, Dover Inc. are delighted to offer their Hunter Electro-Optic Sub System (HEOSS).	
3. Proposal 'theme' – essence of the technical solution	Essence/summation of the solution (desirable)	The refurbishment is designed to restore the system to full operational performance standard, and to make it supportable for a further eight years.	
	Highlighting main selling point(s)/benefit(s) through early mention (optional)	The proposal provides a firm price and programme for the refurbishment of 22 off-Sea Hunter TAGU's for the xxxxx Air Force. SYNCHRO-ELECTRONICS believes this program is an excellent match for its advanced xxxxxxxx product and business plans in terms of performance, price and quantity. Equally, we believe that the excellent working relationship built up between our two companies, even before the merger, during the initial bid phase has demonstrated that working together can be successful.	

Figure 9.2 Generic structure of the executive summary: component realisations

4. Product history	Temporal contextualisation for the product or solution (optional)	**Development Experience in Naval Electro Optical Tracking Systems** The Archer electro-optical gun fire control system was developed in the 1970s and was fitted to the Royal Navy PEACOCK class patrol vessels and exported to customers in the Middle and Far East. .... A further contract was received from the MoD in 1993 to integrate both the electro-optical sub-system and the fire control sub system with the ship's combat system highway.	
5. Product gloss/description	Significant aspects of product design/description glossed. (desirable)	The AB123 has been developed specifically for a low noise missile seeker application and is well suited for use in many other applications which require low noise coupled with high stability. The device is very small and is simple to mount in confined spaces.	
6. Selling point/benefit A	Explanation to Customer about one of the selling points of solution (obligatory)	**Lowest risk** – VSG technology has been demonstrated and is more than just a research programme (in excess of 7,000 units delivered). It is well-established in Military and Commercial markets.	Emphasising the selling point: Low Risk.
7. Selling point/subsequent benefit B	Ditto (optional)	**Strong partnership** The marriage of Inertial Sensor technology from TALC and the manufacturing technology from NNE provides a strong foundation for the supply of Angular Rate Sensors to Davida Technologies.	Emphasising the selling point: Collaboration/Partnership.

Figure 9.2 Continued

Functional component	Discourse function	Example extracts taken from executive summaries in the corpus	Comment
8. Selling point/benefit C, etc.	Ditto (optional)	**Lowest acquisition cost** – Aerospace Group will only fund the difference in cost of developing the IMU from the BOSS-funded IMU development programme.	Emphasising the selling point: Financial benefits.
9. Recommendation/ selling point/benefit	Action(s) and/or product(s) specified (optional)	BOSS recommends the use of an Optical Fire Director (OFD), which has recently been refurbished by BOSS, to replace the existing Forward OFD on xxxxxxxxxx. This has the advantage of saving time and costs.	
10. Closing/conclusion	Final (or reiteration of) particularly impressive selling point (optional)	In addition to providing a cost-effective product, DOVER has a policy of through-life support, for both the equipment and the customer. The system will be fully supported throughout its normal lifespan.	Final sentence or final paragraph, or last sentence in final paragraph. Could be couched as (ostensibly) a recommendation.
	Future reference – encouragement of further dialogue/future action/design devt. potential/next step (optional)	During this program TALC remains ready and willing to assist ZORM in further refining the gyro requirements to meet the important technical and schedule requirement. TALC also welcomes discussions on how the sensor configuration may be modified to meet the long term performance and cost goals of this program.	

Figure 9.2 Continued

Prospective view of reading task, i.e. provision of broad proposal structure (optional)

PROPOSAL FORMAT

This proposal has been written to comply with the format prescribed in the Proposal Preparation Instructions and has been sub-divided into the following volumes:

Volume 1 Management

Volume 1 describes how BOSS will manage the programme of work to meet the contractual commitments.

Volume 2 Technical

Volume 2 (this volume) describes the manufacturing and quality approach which will be adopted to comply with the requirements of the Statement of Work.

Volume 3 Risk

Volume 3 contains the risk response.

Volume 4 Cost

Volume 4 contains the commercial response and pricing details.

Figure 9.2 Continued

(a corpus of 29 executive summaries) to provide an idealised view of their generic structure. Judging by the data, only three components appear to be obligatory, because they feature in every summary in the corpus, and are therefore labelled as such. However, in view of its purpose, it could be said that every executive summary should include an encapsulation of the proposal theme. Furthermore, where compliance is an issue, a comment on the degree of compliance would seem a sensible inclusion. In the diagram, these components have been labelled as 'desirable', to show that they ought to be included, even though the data shows that they are (often) omitted. This may be due to the lack of time available for writing the executive summary for many proposals, resulting in such oversights.

Figure 9.2 shows a generic structure for engineering executive summaries, and includes examples of actual summary components. The left-hand column contains the spine outline structure shown in Figure 9.1, which is supported by examples of components in the column to the right. It can be seen that components may be realised as clauses, sentences, or paragraphs, and that these may be flagged by a heading or subheading. It may also be the case (on rare occasions) that a sentence may contain more than one information component, as in the following example:

The system architecture is based around standard PC hardware and commercial off-the-shelf software [Product Gloss], which ensures a highly cost effective [Benefit] and low-risk solution [Benefit].

Some names in the examples given have been changed or obliterated for companies and products to remain anonymous.

As Chapter 8 explains, there are different categories of proposal, determined by several factors, significant ones being whether the proposal is competitive or non-competitive, solicited or unsolicited, and whether the company is the prime contractor or a sub-contractor. It follows, then, that there should be a variety of summaries in terms of presentation, format, structure, and information content to accompany them, as the following sections show.

9.2.4   Examples of executive summaries

The texts shown below portray a mix of proposal types. The first two are complete texts and the last two are extracts, being complete pages from two longer executive summaries. This selection portrays different modes of presentation used to persuade the chief readers (or chief executives) to agree to, or select, the proposal. All have been produced by a team of writers, with the exception of the first summary, which was drafted by an engineer and then checked by his colleagues.

They reflect presentational and stylistic variation ranging from a more conservative and traditional approach to one incorporating desktop publishing features and electronic formats. The range shows a contrast between a short monochromatic summary presented in a plain format (Figure 9.3) at one end, to a 16-page summary arranged in columns using colour, pictures, and other embellishments (Figure 9.6) at the other. These presentational differences reflect the changes that have taken place in proposal writing over the last 15 years. Each example is accompanied by information about the summary, for example physical proportions, format, presentation, and the extent to which the summary is representative.

Example 1 – a traditional summary without embellishment

This is a whole executive summary, much resembling business abstracts to be found in other (commercial) fields. It is short (211 words), concise, occupies half a page, and comprises plain unembellished text, with no sub-headings, visuals, or other presentational enhancements. It is by no means the shortest executive summary to have been written, as there are shorter ones to be found in the corpus, the shortest comprising 153 words. Engineers aim to produce concise, clear writing and the executive summary is a genre that ideally suits this purpose. However,

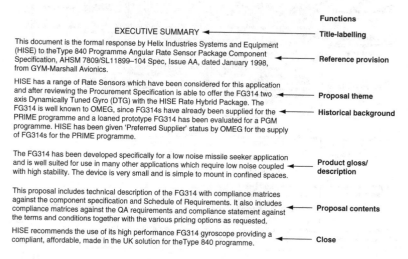

Figure 9.3 Executive summary, example 1 – traditional plain format

some engineers may be over-zealous in their attempts at brevity, putting this above the need to communicate clearly.

Discourse functions performed by different parts of the text are shown in the right-hand margin. These are expanded on later in the chapter, where the communicative function and discourse features of proposals and executive summaries are compared.

Example 2 – a summary of the main benefits of a proposal

Figure 9.4 provides a view of a whole-page executive summary (375 words).

Presentation and format: Figure 9.4 shows the summary's original appearance and structure, which comprises benefits (or selling

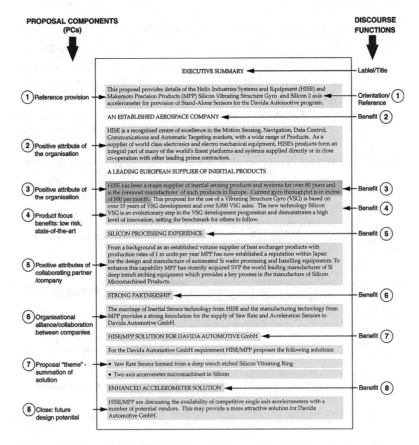

Figure 9.4 Executive summary, example 2 – presented to show key benefits

points) organised under, and highlighted by, sub-headings, which are emphasised in the text through the use of capital letters in a bold font. This type of presentation is a popular format with more than half the summaries in the corpus being in this style, and the length is fairly representative, since most of the summaries are between one and two A4-pages in length.

Discourse functions: This summary is also presented with annotations and shading, to provide an overview of its discourse structure. Segments of text representing particular discourse functions, outlined in Figure 9.2, are now shown, with functional labels for the discourse functions shown in the left-hand margin and details of particular PCs on the right.

It can be seen that the executive summary comprises eight distinct components, realised as segments of text, six of which are benefits, expressed in overtly persuasive language, and two (1 and 7) which are not. No. 7, for example, is a summation of the overall solution, a straightforward proposal statement (i.e., 'OMI/LPP proposes the following') followed by the naming of the two complementary products the engineers wish to 'sell'. By contrast, No. 6 is clearly a bona fide 'benefit' being put forward as part of the proposed solution. Although a short segment, No. 6 contains examples of clearly persuasive elements. The heading, announcing the benefit of a collaborative effort, juxtaposes 'strong' with 'partnership', respectively, an adjective with positive connotations and a similarly positively loaded noun. There follows immediate reinforcement of the concept of a strong partnership through the use of 'marriage', which in a general sense is positively loaded and conveys the notion, again, of holding hands, working closely together, and getting on together. Further evidence of persuasive language lies in the reiteration of 'technology' in 'Inertial Sensor technology from OMI' and 'manufacturing technology from LPP', 'strong' in 'strong foundation' (another juxtaposing of words in a positive collocate).

Example 3 – mimicking newspaper format to present benefits

The third example (Figure 9.5) shows how attempts have been made to present the benefits in a more informal, reader-friendly format. Figure 9.5 shows the whole of the last page of an executive summary, which in totality comprises a total of 1662 words over 11 pages. In this extract, the selling points are organised under red headings and arranged in two columns in the style of a newspaper article or in-house

The sound design approach, extensive production experience and specialised facilities developed at HISE ensure the best technical solution, while our concentration on supply chain management and continual process review assures reliable delivery dates.

PROGRAMME MANAGEMENT

Project teams within HISE are fully experienced across a wide range of national and international projects involving the supply of military and commercial equipment. Examples included:

- Archer 30 APOD System for MoD (RN) (£20M)

- Line Autotrack System for Gen Marshall Corporation (£10M)

- Laser Inertial Navigation Units for TH302 helicopter (£10M)

- Countermeasure Systems for Osborne Inc. (£10M)

- Fire Control Elevation Platforms for Osborne Inc. (£4M)

- TERPROM Ground Proximity Warning Systems for Longue Madison (£multimillion)

All of these were successfully concluded using a dedicated Project Management team to control and lead the programmes.

ESTABLISHED SUPPORT

Support proposals for LATIS are derived from the comprehensive world-wide support service for APOD, which provides a full spares, repairs and post-design services infrastructure to ensure continuing effectiveness of the system throughout its service life.

- Full support infrastructure is already in place

- Proven spares, repairs and post design services

- Highly experienced installation, setting to work and trials team

- LATIS reference rig will be maintained throughout the systems life

These support arrangements can be replicated through partners in the participating nations.

WORKSHARE

Workshare arrangements can be provided in France and Italy to any reasonable level required by the AHSM. We have concluded outline agreements with MARDER SA Defence and Security Division in ▓▓▓▓, and Firenze-Arezzo in ▓▓▓▓. Workshare could include procurement of material, sub-assembly manufacture and support tasks, installation, setting to work, trials, documentation, training and in-country support to the requirement of the nations.

VALUE FOR MONEY

LATIS captures the experience of designing, producing and supporting 24 APOD systems and offers the low risk and assured performance of a Non Developmental Item. Appropriate technology insertion includes state of the art sensors for excellent detection and tracking performance, and COTS electronics to reduce acquisition costs and improve through life supportability. Our experience of manufacturing and overhauling the sensor head has produced a mature and very low risk solution to the major design challenge of the stabilised head.

The result is a system which meets the performance and ARM requirements for TIS, offers assured through life costs, and minimises programme risks. The application of value engineering while deriving LATIS from APOD has ensured that the benefits of LATIS are offered at a competitive price, ensuring a value for money solution to the TIS requirement.

Executive Summary
EP1234

Figure 9.5 Executive summary, example 3 – mimicking popular publications

magazine, in an attempt to impress the reader with, to use a technical author's words, 'a more snazzy and readable layout'. It has already been mentioned that the preferred term for these selling points is 'benefits', and that the engineers try to express these using an overtly persuasive style. 'ESTABLISHED SUPPORT' and 'VALUE FOR MONEY' are clearly signalled benefits, although the heading 'PROGRAMME

MANAGEMENT' conveys a message that is not so obviously beneficial. This segment ostensibly provides information about the proposal team's programme management experience, but is in fact listing this to impress the reader about the company's track record with the product in question. The term 'pedigree', which has been coined by engineers to express this notion of track record or accumulated expertise, has been adopted, and serves here as a label for this particular benefit.

Example 4 – in the style of a glossy magazine

This example reflects engineers' awareness that they need to more consciously 'sell' their products. There are occasions with large proposals (worth many millions of pounds) when engineers communicate directly with the customer, usually restricted to formal face-to-face presentations to the customer's team after the proposal has been submitted (see Chapter 7). Example 4 was composed with the oral presentation in mind. Part of a proposal consisting of several volumes, this executive summary comprises 2988 words, numerous diagrams, and pictures, arranged over 24 pages. The realisation that the chief readers may rely solely on the executive summary to make a decision motivated the project team to include more information about the proposal in the summary itself, and to present it in, what the engineers perceive to be, a 'reader-friendly' format. It is common with proposals of this size to put more information into the executive summaries, and to use pictures and other graphical representation liberally to portray test and statistical information in order to interest the readers, and, ultimately of course, to impress them. The extract in Figure 9.6 shows headings used to specify particular benefits, incorporated into a split-page layout. In this particular case, the main body text appears on the left, whilst 'side-text' (sometimes called 'sidelines') is placed in the right-hand column. Much like advertising slogans, the engineers hope they will be memorable. These slogans attempt to encapsulate in a few words the benefits that they juxtapose.

9.2.5 Executive summaries and proposals: a structural comparison

This section discusses the results of a small case study, centred on the analysis of five randomly selected technical proposals and accompanying executive summaries, viewed as separate texts. The proposal for one of the bids did not include an executive summary, so the collection of summaries contains one that is 'unattached'. The study yields a few

INTEGRATED LOGISTIC SUPPORT

YGO46 All General Purpose Electro-Optical Director

HISE has supplied 24 APODs to Type 22 and Type 23 frigates. Through this contract HISE has Valuable experience in the specification, design and integration of electro-optical surveillance and tracking systems for the naval environment and maintains an experienced support team familiar with installation, trials and through life support of systems for the Royal Navy.

Specification, design and integration in the naval environment

Established support infrastructure

Support proposals for EOSS are derived from our experience in the comprehensive world-wide support of YGO46, which provides a full spares, repairs and post-design services infrastructure to ensure continuing effectiveness of the system throughout its service life.

- Full Royal Navy support infrastructure is already in place

- Proven spares, repairs and post-design services capability

- Highly experienced installation, setting to work and trials teams

- Experienced in the management of Thermal Imager support using a major sub contractor

Proven Royal Navy support Infrastructure

THROUGH LIFE COSTS

The major Through Life Cost driver is the thermal imager. The Longue Madison NIN has the significant advantage of a focal plane array detector, with the reliability benefits due to the elimination of a mechanical scanner. This gives a 20% MTBF gain over the imagers using a scanned long linear array. There are five line replaceable items included in the Longue Madison imager:

- Control Processor Electronic Card Assembly

- Integrated Detector Cooler Assembly

- Power Supply and Focus Control Electronic Card Assembly

- Chassis assembly

- Wiper Motor

Helix Industries
Systems & Equipment

EP1234 Executive Summary   Page 12

Figure 9.6   Executive summary, example 4 – in the style of a 'glossy' magazine

observations that might be useful to those involved in teaching about or writing these texts. The text in the proposals and summaries was analysed to identify the different types of information they contained and the discourse functions of different parts of each document. Put

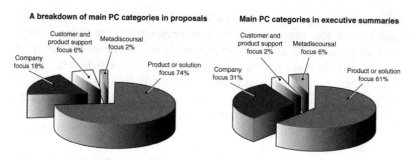

Figure 9.7 Main PC categories in executive summaries

simply, the PCs described in Chapter 8 were identified in a sample of texts and certain aspects examined. Figure 9.7 shows the textual extent in the sample, made up of the technical sections of five proposals and five executive summaries, of the four main PC categories:

1. product/solution-focused
2. company-focused
3. customer support-focused
4. metadiscoursal focused.

The results are summarised graphically as pie-charts, providing a broad view of text 'allocated' to each. A more detailed breakdown of all the PCs is shown later in Figures 9.8 and 9.9. First, however, let us consider Figure 9.7. The two pie-charts are juxtaposed to provide a simple graphical view of the four main PC groups. These show proportions of text relating to each of the main PC categories in the technical proposals and executive summaries, representing figures derived from raw character counts of the texts. These figures were obtained through the use of qualitative data analysis software (NUD*IST Vivo).

Comparing the two, it can be seen that the largest PC category in both relates to information about the product, with 74 per cent and 61 per cent of the total text being devoted to product-focused PCs in technical proposals and executive summaries respectively. These high proportions are to be expected, since they describe the product to the potential customer. These proportions may also be significant for another reason: the technical description of the product in a proposal may be the start of a textual chain of document production, all concerned with specifying the product. It has been suggested that particular documents perform a special role as quasi-products or substitutes for products (see Section 3.2.2). The description of the product in the proposal is the first occurrence in a line of such textual product-substitutes that the engineer writes in the form of documents. A company receives stage-payments when these documents are presented to the customer, until the product itself is delivered, and final payment is made. It is also logical that the largest portion of the technical proposal should be product-focused, since it is this aspect which engages (in all senses of the word) design engineers most.

The second largest PC category in both executive summaries and technical proposals relates to information about the company (see Chapter 8 for more information on company-related PCs). The proportion is large, forming up to 31 per cent of the executive summaries and 18 per cent of the technical proposals. This shows how, in any type of proposal, a company is selling itself as well as the product and, furthermore, that there is a need for the company to 'sell' itself as part of the solution being proposed. Clearly, it needs to do more of this for customers in countries where it is less well known, but less to established customers, like defence ministries and departments, or companies with which it already has a business relationship.

9.2.6   Patterns of information structure

Figures 9.8 and 9.9 show PC patterns in the 10 texts, each represented by a horizontal bar, the total length of which represents the whole of the text, be it the technical proposal or the executive summary. PCs in these diagrams are arranged in order of their appearance in the text, their lengths proportionate to the amount of text devoted to them in the original text. Each bar provides a graphic representation of:

1. the PC membership of each text
2. the amount of text used to express each PC
3. patterns of PC ordering.

It can be seen that Proposal No. 1 and Executive Summary No. 1 mirror each other rather well, both reflecting a balance of company- and product-related PCs. Product solution-focused PCs are clearly the most important, since the function of around 30 per cent of total text relates to just one such PC, the product (or solution) gloss.

However, clear differences between executive summaries and proposals are evident in both Figures 9.8 and 9.9, for example the larger

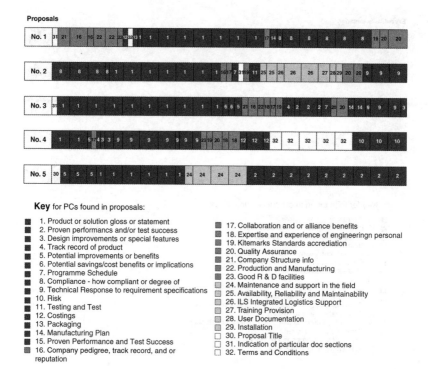

Figure 9.8 PC patterns in technical proposals

number of PCs in technical proposals, amounting to a total of 34, compared to half that number for the executive summaries.

The customer's writing requirements: creativity versus constraint

Legalities and the customer's original writing specification account for the rather odd order of appearance of two sections, one on Terms and Conditions and the other concerned with Risk, and the fact that they comprise such a large proportion of the proposal. This last point is particularly significant when looking at the pattern of PCs, because it reflects constraints imposed by the customer, who has a big influence on the structure of proposals submitted. It can be seen that PC No. 1, a gloss/technical description of the product, appears first, or near the beginning, in all the proposals. In two of the proposals, product compliance and potential improvements are also given early prominence. These early placed product PCs are the result of a creative kind

Executive Summaries

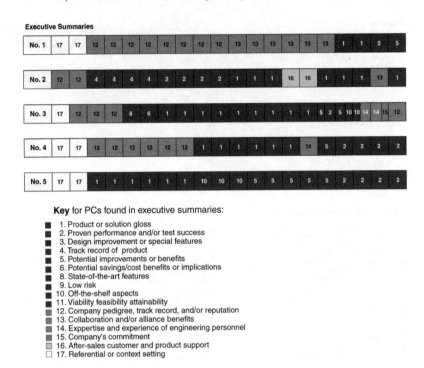

Key for PCs found in executive summaries:

- 1. Product or solution gloss
- 2. Proven performance and/or test success
- 3. Design improvement or special features
- 4. Track record of product
- 5. Potential improvements or benefits
- 6. Potential savings/cost benefits or implications
- 8. State-of-the-art features
- 9. Low risk
- 10. Off-the-shelf aspects
- 11. Viability feasibility attainability
- 12. Company pedigree, track record, and/or reputation
- 13. Collaboration and/or alliance benefits
- 14. Exppertise and experience of engineering personnel
- 15. Company's commitment
- 16. After-sales customer and product support
- 17. Referential or context setting

Figure 9.9   PC patterns in executive summaries

of technical writing produced by engineers as a result of brainstorming together and deciding how best to construct a persuasive document. These are very different in nature from the PCs at the end, which are also product/solution-related, but are usually relegated to a formal (often tick-box format) technical response in appendices, for example degree of compliance, risk, testing and test results, and quality assurance.

The fact is, of course, that the 'solution' is the whole of the proposal, and includes a host of these other considerations underpinning or impacting on the engineering design. It has to be said, though, that these later sections are rather pedestrian to write, and to read. What the customer wants the customer must get, but these customer-imposed requirements may account for the distinctly less persuasive language that is used to express these later sections of the proposal. Nevertheless, engineers try to be persuasive even in these bureaucratically inspired sections, as described in Chapter 2. It is clear that, without this kind interference, proposals would be structured very differently indeed, to be distinct from each other and recognisably rhetorical and persuasive from cover to cover.

Product-focused PCs receive the most attention

Product-focused PCs are naturally obligatory, and would be expected to form the bulk of text in technical proposals and executive summaries, as evidenced by the data and common sense, the exception of Executive Summary No. 1 notwithstanding. Executive summaries, as a rule, tend to be more heavily product-focused (e.g., Executive Summary No. 2), with Executive Summary No. 5 being an extreme example, since it includes not a single mention of the company or the engineers. The executive summary is supposed to present the main benefits of the product (or 'solution') being proposed, and we would expect a substantial proportion to comprise such PCs. Our expectation that many of the PCs in the summary would be clearly persuasive is fulfilled by the results of the analysis that show evidence of PCs such as 'proven performance', 'state-of-the-art features', and 'low risk'. These receive more detailed discussion in earlier chapters.

Company-focused PCs prominently placed

It can be seen that four out of the five executive summaries begin with a metadiscoursal PC, although these are usually headings and, as such, serve to lead readers to the first-mentioned persuasive item, company-related information. It would seem such prominence in the listing accords this information high value, as it does information about the product, and the following order of appearance appears to be a general pattern:

COMPANY/PERSONNEL INFORMATION + PRODUCT INFORMATION

This is understandable, since, as one engineer explained: 'You need to establish your credentials before you can expect them to believe what you say.' In the case of Executive Summary No. 1, most of the executive summary is concerned with company-related information, although this is not reflected (or supported) in the main technical proposal, which is more concerned with describing the product and explaining how well it complies with the customer's requirement. In the case of this proposal, it would have been reasonable to expect the product to feature more prominently in the summary, especially as the executive summary is supposed to encapsulate the main benefits of the solution being proposed. Furthermore, clearly persuasive PCs that receive prominence in the executive summaries do not receive the same treatment in proposals, for example, persuading the reader about the long pedigree and good reputation of

the company. Other PCs, like proven performance, design improvements, or special features, track record of the product, and potential
improvements or benefits are not reflected to the same extent in the
technical proposal, where they either receive markedly less prominence,
or, in most cases, appear to lack any supporting discussion at all. These
observations raise questions about the mismatch. When this happens,
there are three possible reasons:

1. The proposal team have a freer rein when writing the executive
   summary, and made use of this relative freedom to write about special
   features of the solution that could not be included in the technical
   proposal, because of writing constraints (imposed by the customer).
2. The executive summary was written in haste, almost as an after-
   thought in the very last stages of proposal preparation, with whoever
   was tasked with writing it paying scant heed to the main selling
   themes of the proposal.
3. The intended audience is different, and may be managers and not
   engineers.

Customer and product support-focused PCs

In the executive summary sample, there is only cursory mention of
information about after-sales support, more specifically:

Comprehensive support is provided to the RN and MIM export
customers, with a full spares, repairs and post-design services infrastructure to ensure the continuing effectiveness of MIM systems
throughout their practical life.

This dearth is somewhat surprising, for four reasons:

1. Customer and Product Support is often touted within the industry as
   being integral to the successful use of most systems and products. It is
   also received wisdom that it should be considered by design engineers
   from the outset of designing a product, and should be a significant
   feature of any proposal.
2. Substantial time, money, and engineer effort is invested in preparing
   the sections on Product and Customer support, also referred to as ILS,
   in the main part of the proposal.
3. Customers have stated categorically that this aspect is of fundamental
   importance, and influential in the consideration of bids; the customer

is concerned as much about what happens when the product is in use as the design of it. In spite of the acknowledged importance of ILS, however, in reality, ILS tends to be a cinderella domain in proposal and executive summary writing, with less generous staff time and resourcing allocated to it. It is possible that the dynamic process of proposal writing channels energies into other more immediate aspects of the proposal. (Other reasons are also suggested in the discussion of writing teams' disproportionate membership in Section 6.1.5.) This may account for the writing specifications, often issued by the customer, that ensure the neglect of these PCs in executive summaries is not reflected in the main part of the proposal.

4. If the truth be told, design engineers consider ILS worthy but not that interesting. They refer to ILS features as 'hygiene factors'. As one engineer put it, 'you have to have some, but no one wants any more than necessary'.

Metadiscoursal PCs locate text and orientate readers

Metadiscoursal PCs perform functions, like labelling segments of text, that help orientate readers and influence the manner in which the document is read. All the other PCs, however, are persuasive, often overtly so. This Referential or context-setting PC is realised in four of the five executive summaries in the sample, and in every case comprises a single sentence in the simple present or simple progressive tense. In grammatical terms, all four realisations take the form of declarative sentences following an SVO (+O) A structure (where S refers to 'subject', V to 'verb', O to 'object', and A to 'adverbial'):

This response[S] provides [V] details of the DUNKELD SYSTEMS (Scotland) proposal to develop and supply the Next Generation Rate Sensor Unit (NGRSU) [O] for the Aiming Unit of the Starstreak Missile produced by the Crieff Missile Systems Limited [A].

It is logical, and unremarkable, that executive summaries should begin with a metadiscoursal PC, since they name the main participants in the bid and provide a point of reference for both readers and writers. This PC simply names the project that it represents (often in the form of a heading), and tells readers exactly which executive summary they are reading. It performs a function which is part administrative and part discourse-organising, which is useful for the reader if he or she is reading several executive summaries from different batches of submissions, across a range of different projects, and all in the same reading session.

9.3 Summary

Engineers' main concern is to produce a creative 'solution' in the form of a design, and when they write about it in the proposal they need to persuade the customer that it is the best of all the proposals put to him. No one assumes the chief executive reads the whole proposal, let alone sections of it, since a proposal can be a long document, comprising several volumes. The executive summary provides engineers with the opportunity of encapsulating the proposal for him to ponder over. By crystallising their 'solution' in this way, engineers have the opportunity to highlight its benefits, which would otherwise be lost in the detail of the main proposal documents. The intellectual demands of summarising a complex document like an engineering proposal make executive summaries difficult to produce, bringing more anxiety to an already demanding writing scenario. It would seem that this factor, together with engineers' lack of knowledge about how the customer reads it or regards it, contributes to it often being left until the final hurried stages of the bid process when time is short. This is a situation that should be avoided, as the executive summary can be a key document in the bid process. It would be a mistake to underestimate its role in the customer's decision-making.

References

American Department of Defence (1994 and 1998) *Software Development and Documentation* – MILL STD 498 and 2167A.

Association Europeenne Des Constructeurs De Materiel Aerospatial (1989) *AECMA Simplified English: A Guide for the Preparation of Aircraft Maintenance Documentation in the International Aerospace Maintenance Language*. AECMA Document: PSC-85-16598.

Austin, J.L. (1975) *How to do Things with Words*. Cambridge, Mass.: Harvard University Press.

Austin, M. (1990) *The ISTC Handbook of Technical Writing and Publication Techniques*. Slough, UK: Institute of Scientific and Technical Communicators.

Bargiela-Chiappini, F. and C. Nickerson (eds) (1999) *Writing Business: Genres Media and Discourses*. Harlow, UK: Pearson Education Ltd.

Bazerman, C. and J. Paradis (eds) (1991) *Textual Dynamics of the Professions: Historical and Contemporary Studies of Writing in Professional Communities*. Wisconsin, London: The University of Wisconsin Press.

Berkenkotter, C. and T. Huckin (1995) *Genre Knowledge in Disciplinary Communication*. Hillsdale, USA: Lawrence Erlbaum Associates.

Bhatia, V. (1993) *Analysing Genre – Language Use in Professional Settings*. Harlow, UK: Longman Group UK Ltd.

Bolinger, D. (1980) *Language the Loaded Weapon*. Harlow, UK: Longman Group Ltd.

Brazil, D. (1985) *The Communicative Value of Intonation in English*. English Language Research, University of Birmingham, UK.

Brookes, T. (2005) Private communication on engineering design processes.

Brooks, C. and R.P. Warren (1952) *Fundamentals of Good Writing*. London: Dennis Dobson. In Urquhart, A.H. and C. Weir (1998) *Reading in a Second Language: Process, Product and Practice*. Harlow, UK: Addison Wesley Longman Ltd.

Brown, G. and G. Yule (1983) *Discourse Analysis*. Cambridge, UK: Cambrideg University Press.

Brusaw, C.T., G.J. Alred and W.E. Oliu (1976) *Handbook of Technical Writing*. New York: St.Martin's Press, Inc.

Channell, J. (1994) *Vague Language*. Oxford: Oxford University Press.

Chen, H., J. Nunamaker, R. Orwig and O. Titkova (1998) 'Information visualization for collaborative computing', in 'Computer', 31(8): 75–82, IEEE Computer Society Press.

Coulthard, R.M. (ed.) (1994) *Advances in Written Text Analysis*, London, UK: Routledge.

Covey, F. (1997) *Style Guide – For Business and Technical Communication*. 3rd Edition. Salt Lake City, USA: Franklin Covey Company.

Davies, F. (1995) *Introducing Reading*. London, England: The Penguin Group.

Davies, F. and G. Forey (1995) *Effective Writing for Management Project*. School of Education. Bristol University.

Davies, F. and T. Greene (1984) *Reading for Learning in the Sciences*. Schools Council Publication. Edinburgh: Oliver & Boyd.

Davies, F., G. Forey and D. Hyatt (1999) 'Exploring aspects of context: selected findings from the Effective Writing for Management project'. In Bargiela-Chiappini, F. and C. Nickerson (eds) (1999) *Writing Business: Genres Media and Discourses*. Harlow, UK: Pearson Education Ltd.

Dependable Computing Systems Centre (DCSC) (1996) University of York and the University of Newcastle, United Kingdom. General information publicity brochure.

Devitt, A.J. (1991) 'Intertextuality in tax accounting: Generic, referential, and functional'. In Bazerman, C. and J. Paradis (eds) (1991) *Textual Dynamics of the Professions: Historical and Contemporary Studies of Writing in Professional Communities*. Wisconsin, London: The University of Wisconsin Press.

Dobrin, D.N. (1989) *Writing and Technique*. Urbana, Illinois: National Council of Teachers of English.

Dudley-Evans, A. (1986) 'Genre Analysis: An investigation of the introduction and discussion Sections of MSC Dissertations'. In Coulthard, R.M. (ed.) (1994) *Advances in Written Text Analysis*, London, UK: Routledge.

Ellis, R. (1997) *Communication for Engineers*. London: Arnold.

Fear, D.E. (1977) *Technical Communication*. Scott, Foresman and Company, Illinois: Glenview.

Fitzgerald, J.S. (1993) 'Formally Specifying a Trusted Gateway'. Unpublished paper. Department of Computing Science, University of Newcastle-upon-Tyne.

Freed, R.C. (1987) 'A mediation of proposals and their backgrounds'. *Journal of Technical Writing and Communication*, 17(2). Baywood Publishing Company Inc.

Grabe, W. and R. Kaplan (1996) *Theory and Practice of Writing*. London and New York: Addison Wesley Longman Ltd.

Halliday, M.A.K. (1978) *Language as a Social Semiotic*. London: Edward Arnold.

Halliday, M.A.K. (1994) *Functional Grammar*. 2nd Edition. London: Edward Arnold.

Halliday, M.A.K. (2004) *The Language of Science*. London, UK and New York: Continuum.

Halliday, M.A.K and J. Martin (1993) *Writing Science – Literacy and Discursive Power*. London: The Falmer Press.

Haslam, J.M. (1988) *Writing Engineering Specifications*. London/New York: E. & F.N. Spon Ltd.

Heffer, C. and H. Sauntson (eds) (2001) *Words in Context: A Tribute to John Sinclair on his Retirement*. University of Birmingham: English Language Research Discourse Analysis Monograph No.18.

Hicks, T.G. (1961) *Writing for Engineering and Science*. New York: McGraw-Hill Book Company, Inc.

Hoey, M. (2001) *Textual Interaction – An Introduction to Written Discourse Analysis*. London and New York: Routledge.

Hopkins, A. and A. Dudley-Evans (1988) 'A genre-based investigation of the discussion sections in articles and dissertations'. *English for Specific Purposes*, 14: 115–126.

Houp, K.W. and T.E. Pearsall (1980) *Reporting Technical Information*. 4th Edition. Glencoe Publishing Co. Inc.

Huckin, T.N. and L.A. Olsen (1983) *English for Science and Technology*. McGraw-Hill Book Company.

Hunston, S. and G. Thompson (2000) *Evaluation in Text*. Oxford: Oxford University Press.

ITSEC (1992) – *Information Technology Security Evaluation Criteria*. UK IT Security Evaluation and Certification Scheme. HMSO Books (PC16).

James, L. (1997) 'Providing Pragmatic Advice on how Good your Requirements are – the Precept "Requirements Counsellor" Utility'. Unpublished paper. Integrated Chipware Limited.

Jones, J.V. (1989) *Logistic Support Analysis Handbook*. TAB Professional and Reference Books. Blue Ridge Summit, PA, USA: McGraw-Hill Inc.

Jones, J.V. (1995) *Integrated Logistics Support*. 2nd Edition. New York: McGraw-Hill Book Co., Inc.

Kidd, C. (2001) 'The case for configuration management'. *IEE (Institution of Electrical Engineers) Review*, September 2001.

Kincaid, R. (1997) *A Dinosaur in Whitehall*. London, UK: Brassey's.

Kirkman, J. (1992) *Good Style-Writing for Science and Technology*. London, UK: E. & F.N. Spon.

Kress, G. and T. van Leeuwen (1996) *Reading Images – The Grammar of Visual Design*. London, UK: Routledge.

Latour, B. and S. Woolgar (1986) *Laboratory Life; The Construction of Scientific Facts*. 2nd Edition. Princeton: Princeton University Press.

Marder, D. (1960) *The Craft of Technical Writing*. New York: The Macmillan Company.

Marshall, H. (1986) Quantity Surveying Reports, M.A. Dissertation. University of Birmingham, UK.

Martin, J.R. (1989) *Factual Writing: Exploring and Challenging Social Reality*. Oxford, UK: Oxford University Press.

Meyer, B. (1985) 'On Formalism in Specifications'. Institute of Electrical and Electronic Engineers (Software).

Myers, G. (1990) *Writing Biology – Texts in the Social Construction of Scientific Knowledge*. Madison, Wisconsin: The University of Wisconsin Press.

Newell, M. (2005) Editorial, Communicator. *Journal of the Institute of Scientific and Technical Communicators*. Winter 2005.

Newman, L. (2003) *Proposal Guide – For Business Development and Sales Professionals*. 2nd Edition. Farmington, UT, USA: Shipley Associates.

O'Brien, S. (2005) 'Controlling your language', *Communicator: Journal of the Institute of Scientific and Technical Communicators*. Autumn 2005.

Ochs, E. and S. Jacoby (1997) 'Down to the Wire: The cultural Clock of Physicists and the Discourse of Consensus'. *Language in Society*, 26(4), 479–505.

Odell, L. and D. Goswami (eds) (1985) *Writing in a Nonacademic Setting*. New York: The Guilford Press.

Packard, V. (1957) *The Hidden Persuaders*. London, UK: Longman.

Paradis, J., D. Dobrin and R. Miller (1985) 'Writing at Exxon ITD: Notes on the writing environment of an R&D organization'. In Odell, L. and D. Goswami (eds) *Writing in Non-Academic Settings*, New York: The Guilford Press.

Pauley, S. (1973) *Technical Report Writing Today*. Boston, USA: Houghton Mifflin Company.

Pike, K. (1981) 'Grammar versus reference in the analysis of discourse'. *Tagmemics, Discourse and Verbal Art*. Michigan: University of Michigan Press. In Hoey, M. (2001) *Textual Interaction – An Introduction to Written Discourse Analysis*. London and New York: Routledge.

Quirk, R. and S. Greenbaum (1973) *A University Grammar of English*. London. UK: Longman Group Ltd.

Quirk, R., S. Greenbaum, G. Leech and J. Svartvik (1985) *A Comprehensive Grammer of the English Language*. Harlow, England: Longman Group Ltd.

Riddle, S. and A. Saeed (1998) 'Tracking Conflicting Requirements and Trade-Offs'. Research paper. BAe Dependable Computing Systems Centre/Department of Computing Science, University of Newcastle-upon-Tyne.

Sales, H.E. (2000) 'Moody modals: (mis)interpretations of "shall" and "will" in engineering specifications'. In Heffer, C. and H. Sauntson (eds) (2001) *Words in Context: A Tribute to John Sinclair on his Retirement*. University of Birmingham: English Language Research Discourse Analysis Monograph No.18.

Sales, H.E. (2002) *Engineering Texts: A Study of a Community of Aerospace Engineers, their Writing Practices and Technical Proposals*. PhD Thesis, University of Birmingham, United Kingdom.

Saville-Troike, M. (1989) *The Ethnography of Communication*. 2nd Edition. Oxford, UK/Malden, Mass, USA: Blackwell Publishers Ltd/Inc.

Searle, J.R. (1969) *Speech Acts – An Essay in the Philosophy of Language*. Cambridge/ the Syndics: Cambridge University Press.

Segal, J. Z. (1993) 'Writing and medicine'. In Spilka, R. (ed.) (1993) *Writing in the Workplace – New Research Perspectives*. Carbondale and Edwardsville: Southern Illinois University Press.

Sinclair, J.M. (ed) (1987) *Looking Up: An Account of the COBUILD Project in Lexical Computing*. London: Collins ELT.

Sinclair, J.M. (2001) 'The deification of information', In Thompson, G. and M. Scott (eds)(2001) *Patterns of Text: In Honour of Michael Hoey*. Amsterdam and Philadelphia: John Benjamins.

Sinclair, J.M. (2004) *Trust the Text*. London, UK: Routledge.

Snow, C.P. (1959) The Two Cultures and the Scientific Revolution. The Rede Lecture 1959. Cambridge, UK: Cambridge University Press.

Snow, C.P. (1964) *The Two Cultures: And A Second Look – An Expanded Version of the Two Cultures and the Scientific Revolution*. Cambridge, UK: Cambridge University Press.

Souther, J.W. (1954) *A Guide to Technical Reporting*. University of Washington.

Souther, J.W. and M.L. White (1977) *Technical Report Writing*. 2nd Edition. New York: John Wiley & Sons.

Spilka, R. (ed.) (1993) *Writing in the Workplace – New Research Perspectives*. Carbondale and Edwardsville: Southern Illinois University Press.

Stross, R.E. (1990) *Preparing Successful Proposals*. Potomac, Maryland, USA: Center for Public Management.

Surma, A. (2005) *Public and Professional Writing: Ethics, Imagination and Rhetoric*. Basingstoke, UK: Palgrave Macmillan.

Swales, J.M. (1996) *Other Floors Other Voices – A Textography of a Small University Building*. Mahwah, New Jersey: Lawrence, Erlbaum Associates, Publishers.

Swales, J. (1981) *Aspects of Article Introductions*. Birmingham, UK: The University of Aston, Language Studies Unit.

Swales, J. (1990) *Genre Analysis – English in Academic and Research Settings*. Cambridge, UK/New York, NJ/Melbourne, Australia: Cambridge University Press.

Swales, J. and C. Feak (1994) *Academic Writing for Graduate Students*. Ann Arbor: The University of Michigan Press.

Texel, P. and C. Williams (1997) *Use Cases Combined with BOOCH, OMT, UML*. Upper Saddle River, NJ: Prentice-Hall Inc.

The British Standards Institution (1998) British Standard 7373: *The Preparation of Specifications*.

Thomas, P. *et al.* (1991) *Software Engineering. Book 1 M860. Software Engineering Study Guide*. Faculty of Mathematics and Computing. The Open University. Open University Press.

Thomas, P., D. Ince, *et al.* (1994) *Computing for Commerce and Industry – Software Engineering*. Book 1. Milton Keynes: The Open University.

Trimble, L. (1985) *English for Science and Technology – A Discourse Approach*. Cambridge, UK: Cambridge University Press.

Urquhart, A.H. and C. Weir (1998) *Reading in a Second Language: Process, Product and Practice*. Harlow, UK: Addison Wesley Longman Ltd.

Van Nostrand, A.D. (1997) *Fundable Knowledge – The Marketing of Defense Technology*. Mahwah, NJ: Lawrence Erlbaum Associates.

Weisman, H.M. (1962) *Basic Technical Writing*. Columbus, Ohio: Charles E. Merrill Books, Inc.

Winsor, D.A. (1996) *Writing Like an Engineer*. Mahwah, NJ: Lawrence Erlbaum Associates.

Index

aboutness, 146, 212
acronyms, 65, 161
    categories of, 176–7
    use of, 174–6; as professional
        display, 178
Alred, G.J., *see* Brusaw, C.T.
American defense industry, formality
    of documents, 18
American Ministry of Defense, 36
American Military standards, 30–1
arts versus sciences, 15–16, 21
audience, 83
    internal and external, 74
Austin, J.L., 95, 138
Availability, Reliability,
    Maintainability, and
    Testability, 42

Bargiela-Chiappini, F., 31
Bazerman, C., 31
benefits in proposals, 80, 127, 240
    in model texts, 219
    selling points, 127, 145,
        158, 220
    'selling' the company, 189, 202
    versus selling points, 189
Berkenkotter, C., 2
bespoke design, 121
    see also product
Bhatia, V., 206
bids, 65, 125
    bid process, 122, 127, 128–30,
        152, 215–17
    'cost-plus' process, 217–18
    see also proposals
boffins, 3
    see also engineers, types of
Bolinger, D., 137, 138, 200
Brazil, D., 185
British Standards, on design, 31
Brookes, T., 56

Brooks, C., 86, 110
Brown, G., 194
Brusaw, C.T., 96

cardinal point specifications, 99
Change Team, 28–9, 106
changes
    attempts to control, 27
    to design, impact of, 111
    implementing, 28–9
Channell, J., 87
Chen, H., 105
chief readers, *see* readers, of proposals
collaborative writing, 36–7, 39–41,
    123–4
commercial off-the-shelf, *see* product
communication breakdown, in
    requirements writing, 111
communication skills
    diagrams, use of, 73
    engineers' aspirations, 19–21
    oral presentations, 178–86
    writing skills, *see* engineers
communication tasks, the four 'P's, 70
community of practice, 22
company procedures, *see* procedures
competitive proposals, *see* proposals,
    types of
compliance, 197
    description of, 79
    matrices, 162
    as proposal theme, 180
configuration management practices,
    163
consequences
    unforeseen: of design work, 24,
        25, 67
    of unsuccessful proposals, 124,
        133–5
controlled languages, 34

copying, as a writing strategy, *see* writing
copyright, *see* proprietary status
Covey, F., 193, 219
creativity
    in engineering, 1, 121
    heralded by straw men, 53
    hindered by prescription, 34
    versus restriction, 14–15
customer
    acceptance of proposal, 65
    descriptions for, 77, 143
    enquiry, response to, 78
    expectations, management of, 68
    fickleness, 121, 132
    influence on engineers' work, 8–9, 27
    meeting the needs of, 56
    oral presentations for, 178–86
    phases of interaction with, 61
    questions, 84
    reaction to engineers' writing, 140
    strategies for persuading: in proposals, 200–5
    support, 69
    writing constraints imposed by, 35–6, 130, 235–6
customer requirement, 45, 98–103, 127, 130–2, 143
    for a document, 45
    needs analysis, 65
    proposal in response to, *see* proposals
    significance of 'shall', 114
    use study, 65

dashes in punctuation, avoiding the use of, 149
Davies, F., 31, 73, 74, 76, 86, 89, 206
dearth of information
    engineering documents, 31–2
    on proposals, 193
Dependable Computing Systems Centre, 25
description
    accuracy of, 146
    general to specific, 101

objective, 18, 136
objective versus subjective, 138
product description, *see* product
    in proposals, 196–201
    of solution in proposals, 92
    sub-system, 110
    system, 78
    technical, 36, 77, 78–80, 86, 91, 92, 95, 108, 150; clause structures in, 88; complex simplicity, 88; modal verbs in, 113–19; persuasive, 123
    topic focused, 208
    use of analogy, 150
    vague, 87, 92
design
    'blue-sky' thoughts about, 78
    British Standards, 31
    consequences: high-level design changes, 111; *see also* unforeseen consequences
    early thoughts about, 57, 59
    freeze, 66
    knowledge capture, 59, 63, 104
    procedures, *see* procedures, and methodologies
    reliability, search for, 24–6
    specificity cline, 61
    subject to scientific proofs, 23, 27
design freeze, *see* design
design process
    changes, 127–8
    knowledge accrual, 56–7, 59
    long time-scales, 60–3, 128, 135, 163
    managing change, 103
    reduction of uncertainty in, 56
    technical notes, 82
    thresholds in, 64
    traceability, 104
    for winning proposal, 127–8
Devitt, A.J., 174
disciplinary community, 22
discourse community, 22
discourse functions, in proposals, 196–201
discourse topics, 196
Dobrin, D.N., 18, 31, 110
document manager, 63

document 'orientation', 63, 64, 65
documents
  for administrative purposes, 76
  aircraft maintenance, 32
  in design, 56–7
  design, main audiences for, 61
  ephemeral, 75, 90
  executive summaries, *see* executive
      summaries
  formality in, 18
  generate income, 93
  log books, 77
  most significant, 74
  problem-causing, *see* problems
      documents
  product-related, 61
  production of, 44
  proposals, 77, 92; *see also* proposals
  for proto-type, 65–6
  reports, 22, 77, 81, 84, 90
  SOFT reports, 74
  specification, *see* specifications
  specificity cline for, 61
  storage of, 81
  straw men, 43
  straw men as provocative, 44
  straw men in production of, 44
  system manual, 45–50
  user-oriented, 69
  written collaboratively, *see* texts
DOORS, 26
Dudley-Evans, A., 206, 207
Dynamic Object-Oriented
    Requirements System, 26

Ellis, R., 88, 96, 193, 194, 195,
    202, 217
Elmer Sperry, 60
engineering
  catastrophes, 103
  methodologies, 24–6, 106
  procedures, *see* procedures, and
      methodologies
  solutions in, 1; *see also* solution
  specifications, 96
engineering talk
  about 'best practice', 68
  about design procedures, 24
  acronyms in, 178

deciding on proposal content, 141
in description, 86
and diagrams, 73
face-to-face with the customer, 84
generating new ideas, 64
in presentations, 180
referring to straw men, 42
straight talking about the solution,
    132
striving for eloquence, 19
talking about the customer, 8
the unmentionable problem, 126
engineers
  attitudes towards school subjects, 3
  controlling work behaviour of, 26,
      27–9, 106
  discomfort over persuasive
      language, *see* persuasion
  displaying their credentials, 152
  finding the right words, 142–51
  involvement in proposal writing,
      129
  as knowledge 'holders', 163
  preoccupation with product, 55
  proprietary about text, 153
  reactions to draft text, 47–50
  reasons for becoming, 3, 5, 6
  reliance on procedures, 23–6
  reluctance to use pronouns, *see*
      writing
  search for design reliability, 24–6
  self-criticism, 19–21
  solution-oriented, 15–16
  types of, 9–13; boffins, 3; female
      engineers, 3
  types of, support, 142
  use of diagrams, 73; *see also* visuals
  use of writing guidance, 31
  versus marketing colleagues, *see*
      persuasion
  write a lot, 6
  writing load, 6
  writing responsibilities, 6
  writing skills, 5, 32
engineers' views on
  communication skills, 19
  ephemeral texts, 73, 76
  factual language, 142
  persuasion, 136–51

software, 26
subjectivity and objectivity, 17–18
text categories, 74
text types, 77
texts, 14–15, 71
texts, prospective, 90
texts, radical, 90
writing proficiency, 5
English
  engineers' distrust of, 26
  'formal', 113
  'natural', 26, 109, 112, 139
  perceived as inadequate, 111
  Simplified, *see* Simplified English
English for Science and Technology,
  189, 206
English for Special Purposes, 90, 189
English for Specific Purposes, 22, 71
ephemeral texts, *see* texts
ethnography, 2
  ethnographic research
    methodology, 2
European Association of Aerospace
  Manufacturers, 32, 33
executive summaries, 84, 145, 155,
  158, 159–61, 214
  discourse functions, 220
  generic outline, 220–6
  outline structure, 226–8
  presentation, 227–31
  purpose, 214–17
  reading, 126–7
  writing under pressure, 216

Feak, C., 188, 206, 219
Fear, D.E., 87, 96, 140
female engineers, *see* engineers,
  types of
financial consequences, *see*
  consequences
Fitzgerald, J.S., 113
Fleet Weapons Acceptance, 69
Forey, G., 31, 74, 76
formality
  tendency towards, 20
  versus informality, 18, 72,
    139, 151
Freed, R.C., 83, 193
functionality, 78, 92

genres, *see under* texts
Goswami, D., 31
government contracts, reliance
  on, 123
Grabe, W., 32, 86, 219
Greenbaum, S., 88
  see also Quirk, R.
Greene, T., 73, 89, 206
gyroscope, 2, 18, 60

Halliday, M.A.K., 73, 88, 89, 90,
  110, 166
Haslam, J.M., 88, 96
Hicks, T.G., 6, 96, 104, 105,
  140, 190
Hoey, M., 72, 207, 208
Houp, K.W., 84, 87, 137, 190,
  191, 192
Huckin, T.N., 2, 179
Hunston, S., 137
Hyatt, D., 31, 76

ideology, in engineering, 1
inertial measurement unit, 132
  see also silicon gyroscope
information density, 89
Information Technology Security
  Evaluation Criteria, 31, 35–6, 116
Institute of Scientific and Technical
  Communicators, 34
Integrated Logistics Support, 42, 130,
  144, 180, 210, 238
  writing load, 130
Invitation to Tender, 42, 109, 194
ISO10007, 163
ITSEC, 35

Jacoby, S., 180
James, L., 105
Jones, J.V., 56, 57, 175

Kaplan, R., 32, 86, 219
Kidd, C., 64, 174
Kincaid, R., 5, 85
Kirkman, J., 18, 35, 95, 96, 114, 140
kite flying, 39, 43
  in text, 43; *see also* straw men
Kress, G., 165, 166

language
  complex noun phrases, 89
  finding the right words, 141–2
  formal and informal, *see* formality
  personal, 151
  scientific, 89
  unambiguous, 88
language structure, complex
      simplicity, 88
Latour, B., 2, 8
Leech, G., *see* Quirk, R.
lexical density, *see* texts
Lightweight Object Repository, 26
logbooks, 27
Lore, 26

Marder, D., 84, 110, 146, 189, 205
Marshall, H.E., 19, 217
Martin, J., 73, 86, 138
metaphors, engineers' use of, *see*
      writing practices
Meyer, B., 105
Miller, R., 31
Ministry of Defence, 175
  relationship with, 123
  standards, 30–1
modal verbs, 36, 110, 113–19
  used by the customer, 102
model texts, 30, 219
  popularity of, 187
  see also texts
Myers, G., 31, 73, 81, 139, 165, 167,
      168, 187

Natural Language Processing, 26
Newell, M., 13
Newman, L., 129, 217, 219
Nickerson, C., 31
non-competitive proposals, *see*
      proposals, types of
NUD*IST Vivo, 233
Nunamaker, J., *see* Chen, H.

objectivity, as a shield, 18
O'Brien, S., 34
Ochs, E., 180
Odell, L., 31
Oliu, W.E., *see* Brusaw, C.T.
Olsen, L.A., 179

oral presentations, 80, 83, 178–86
Orwig, R., *see* Chen, H.

Packard, V., 138
Paradis, J., 31
Pauley, S., 84, 140, 192, 193, 195
Pearsall, T.E., 84, 87, 137, 190,
      191, 192
peer comment, *see* texts; writing
  critical powers on demand, 51
personal versus impersonal, 18
persuasion, 87, 92, 136–51
  disagreement about, 86
  engineers' discomfort about, 122,
      123, 139–42
  engineers versus sales colleagues,
      136
  negative connotations, 138
  as a notion, 139
  restrained, 142–51
  strategies in proposals, *see* proposals
  strategy continuum, 137
  suasion, 138
persuasion strategies, in proposals,
      see proposals
persuasive strategy continuum, 137
Phase Gates, 28–9
philosophical principles, in
      engineering, *see* ideology, in
      engineering
Pike, K., 207
problems
  unforeseen, in design, 67, 103
  with requirements, *see* requirements
      writing
  versus solutions, 16
problems documents, 84–5, 107
procedures
  attempts to control, 121
  company, 27–8
  consultative, 28–9
  and methodologies, 23–6, 53
  often ignored, 30–1, 35
  procurement, 66, 121
  search for improving, 26
  Simplified English, comment on,
      32–5
  tendering, 121
  ticking-off, 29, 66

product
  bespoke, 16, 77
  commercial off-the-shelf, 77, 175
  description, 78–80, 152; of
      appearance, structure
      and construction, 78, 198;
      of performance, 79;
      in proposals, *see*
      proposals; user-oriented, 79
  as dynamic element, 57
  functions of, 78
  life-cycle, 56
  maintenance, 69
  obsolescence, 80
  operation or use of, 67–9, 199
  as solution, 17, 55
  shelf-life of, 60, 80
  specifications, 97
  as textual focus, 77
product ≈ text, *see* texts, as substitute
      for products
product development, 55–6
  asserting control over, 70
  macro-phases, 60–3
  main phases, 61
  main stages, 65–9
product life-cycle, textual
      perspective, 64
professional talk, presentations,
      see oral presentations
pronouns, avoidance of, *see* writing
proposal components, 152
proposals, 81, 84–5, 87, 92
  analysis rationale, 205–12
  benefits or selling points, *see*
      benefits in proposals
  components (PCs), 152, 154, 205,
      206, 208, 210–12, 233
  components taxonomy, 211
  cover illustrations, 163–72
  covers, 162
  discourse functions, 196–201
  discourse topics, 196
  ethical considerations, 126, 133,
      216–17
  executive summary, *see* executive
      summaries
  expensive writing activity,
      123–4, 134

generic, 132, 147, 160–1
glossary, 159–61
guidance for writing, 189–96
information content of, 209
information structure, 234–40
persuasion strategies, 201
product description, 196–201
proprietary statement, 159–61
rapid, straw men, 44
reading, *see* readers
as response to customer
    requirement, 65
'route map', 155–6
seen as technical reports, 190
as selling documents,
    187–8, 193
significant parts, 159–61
structure, 141–2, 158–62
technical section, 159–61
textual cosmetics, 153
textual extent of, 154
themes, 161, 201–5, 203
tight deadlines, 129, 212
title page, 159–61
types of, 124–6; competitive, 65,
    128–30, 192; interfirm, 192;
    intrafirm, 192; letter, 125, 190;
    non-competitive, 65, 128, 192;
    technical, 125
unsuccessful, *see* consequences
proprietary status
  copyright, 160, 172–4
  of technical notes, 82

Quirk, R., 88, 149

readers
  of executive summaries, 216
  of proposals, 126–7; chief readers,
      127, 216, 226, 240; specialist
      readers, 127
  prospective reading, 126
  retrospective reading, 127
report writing, seen as unproblematic,
      84, 90
Request for Information (Request for
      Quotation), 37, 58, 80, 125
  about the silicon gyroscope, 147

Request for a Quotation, *see* Request for Information
requirements, 77, 79, 81, 84–5, 91
  definitions, 93
  hierarchies in, 109
  high-level, 109
  inconsistencies of practice, 117
  lower-level, 109
  (mis)perceptions of English, 112
  modal verbs in, 113–19
  monitoring, 116
  obligatory, 100
  optional, 100
  prime, 100
  software for writing, 26, 106
  and specifications, 26
  see also cardinal point specifications
Requirements Traceability Management, 26, 105, 116, 117
requirements writing, 65
  problems with, 96
research on engineers, data used in, 2
Response to Request for Information, 65, 125, 147
responsibility
  burden of, 23, 103
  widening, for writing, *see* writing
reverse-engineered text, 212
rhetorical community, 22
Riddle, S., 105
RTM, 26

Saeed, A., 105
Sales, H.E., 1, 3, 19, 30, 84, 94, 220
Saville-Troike, M., 2
sciences versus arts, 15–16, 21
scientific approach, 24
Searle, J.R., 188
Segal, J.Z., 207, 208
Segway Human Transporter, 60
selling points, *see* benefits in proposals
shall, 114
  versus will, 115
side-text, 231
sidelines, *see* side-text
silicon gyroscope, 131, 146–51, 164, 176, 209
  see also gyroscope

Simplified English, 31, 32–5
Sinclair, J.M., 26, 90, 206
solution, 129, 240
  compliance of, 79
  contractual obligations regarding, 92
  description of, 144, 196–201
  identification of, 132
  secrecy about, 126, 128
  as product, *see* product
  technical, 130
  versus problem, 16
Souther, J.W., 84, 189, 190, 195
speaking, *see* engineering talk
specifications, 35–6, 65, 77, 84–5, 91, 92
  abstract, 24
  after the event, 67
  categories, 109
  concern about, 94
  contractual, 66
  definitions, 93
  high-level, 65
  incomplete, consequences of, 107
  manufacturing, 81
  modal verbs in, 113–19
  poorly written, impact of, 109
  procedures for writing, 23
  system, 97
  technical hierarchy, 110
  vagueness, 85
  writing guidelines, 97
speech, *see* engineers; professional talk
Spilka, R., 31
straw men, *see* text types
  in academia, *see* text types
  evaluation of, 51–3
  as heralds of creativity, *see* creativity
  as stress-relievers, 52
Stross, R.E., 193, 202, 208
subjectivity versus objectivity, 17–18
Surma, A., 82
Svartvik, J., *see* Quirk, R.
Swales, J., 13, 22, 42, 188, 206, 219
systems design, 26, 95
  software, 26

talking, *see* engineering talk
teams
  bid teams, 122
  reading teams, *see* readers
  team working, 46–50, 51, 53,
    128–30
  team writing, 123–4
technical authors, 13–14, 32, 34, 43,
    47, 152, 153, 156–8, 178
  as textographers, 13
technical proposals, *see* proposals
templates, *see* texts, formats and
    templates for
tendering process, *see* bids
tests
  keeping a record, 27
  and trials of product, 65, 68
Texel, P., 116
text analysis
  analytical framework: for proposals,
    206–12
  topic focused, 208
text genres, 206
text models, 30, 219
  see also texts, formats and templates
    for; model texts
text substitute for product, *see* texts
text types, 81–4
  emails, 82–3
  executive summaries, *see* executive
    summaries
  formal and informal, 72
  of greatest concern, 74
  handbooks, 69, 82
  important to engineers, 22, 77
  lab reports, 32
  letters, 82–3, 85
  logbook entries, 27, 81
  logbooks, 32, 77
  manuals, 45, 69, 82
  manufacturing instructions, 81
  memos, 82–3
  patents, 82
  précis, 219
  procedures, 82
  proposals, *see* proposals
  Requests for Information, 58
  straw men, 41–4
  summaries, 219

technical notes, 82, 84
textographers, 13
  see also technical authors
texts
  aboutness, as textual construct, *see*
    aboutness
  for administrative purposes, 76
  analytical framework for proposals,
    206–12
  annotations on, *see* writing
  categories of, 74
  communicative function, 95
  draft, peer review of, 48
  as dry runs, 52
  as engineered artefacts, 105
  engineers' views about, 71
  ephemeral, 73, 90
  with financial value, 93
  formal and informal, *see* text types
  formats and templates for, 37
  genres, 22
  of greatest concern, *see* text types
  intrinsic value of, 63
  lexically dense, 37
  malleability of, 41
  problem-causing, *see* problems
    documents
  product-focused, 77
  propositional content of, 188
  reports, *see* documents
  reverse-engineered, 212
  as substitute for product, 8, 43, 54,
    63, 172–4
  written collaboratively, 40
themes in proposals, 219
  see also proposals
Thomas, P., 57, 113
Thompson, G., 137
thresholds, 28–9
ticking-off, 29, 66
Titkova, O., *see* Chen, H.
traceability, 162–3
Trimble, L., 110

unforeseen consequences
  in design work, *see* design
  of incomplete specification, 107
Urquhart, A.H., 86

van Leeuwen, T., 166
Van Nostrand, A.D., 18, 37
visuals
    engineers' use of, 163–72
    ideological stance, 165–8
    to impress and influence, 168–70
    'toys for the boys', 171–2

Warren, R.P., 86, 110
Weir, C., 86
Weisman, H.M., 110
White, M.L., 84, 190, 195
Williams, C., 116
Winsor, D.A., 32, 136
women in engineering, dearth of, 4
Woolgar, S., 2, 8
work practices, formal, 41
writing
    annotations of, in peer review, 50
    behaviour, attempts to control,
        27–9
    copying, 19
    definitions, 150
    descriptions, 35–6, 80
    electronic comments, 27
    engineers' writing load, *see*
        engineers
    engineers' writing skills, *see*
        engineers
    executive summaries, *see* executive
        summaries
    facilitative, 83
    lab reports, 32
    lexically dense, 89
    logbooks, 32
    objective, 18, 87, 140–1
    plans, 81

pronouns in, 18
proposals, *see* proposals
proposals, seen as technical
    reporting, 190
reluctance with pronouns, 18
requirements and specifications, *see*
    requirements
resource-hungry activity, 123
responsibilities for, 6–8
risk-averse, 38–9, 75
risk-taking in, 41–4
Simplified English, *see* Simplified
    English
specifications, 23, 35–6, 97
stylistic aspirations, 19, 144, 147
in teams, *see* teams; collaborative
    writing
test procedures, 26
a time consuming activity, 8
widening responsibilities for, 6
writing constraints,
    customer-imposed, 35–6, 235–6,
    238
writing guidelines, *see* writing
    customer-imposed, *see* writing
        constraints, customer-imposed
writing practices
    collaborative, 38–9
    manuals, 45–50
    metaphors used in, 41–4
    outdated, 38–9
    in the British navy, 38–9
writing skills, at undergraduate level,
    32, 81

Yule, G., 194